Encyclopaedia of

Genes, Genomes, Genetics and Chromosomes

Encyclopaedia of

Genes, Genomes, Genetics and Chromosomes

Volume 1

Editor:
Dr. R.N. Shukla

AGROTECH PRESS
JAIPUR ■ NEW DELHI

AGROTECH PRESS

Regd. Off. "Bony Residency", Gate No. 2
Opp. Tilak Public School, Vishweswariya Nagar,
Gopalpura Road, Jaipur - 302018, India
Ph: 0141 - 2761280, Telefax: 2761381
E-mail : agrotechpress@gmail.com

Delhi Office: 102, 1st Floor, Satyam House,
4327/3, Ansari Road, Darya Ganj,
New Delhi - 110002, India
Ph: 011-45652440
visit us: www.agrotechpress.com

Distributed By
KEYA INTERNATIONAL PRESS LLP
E-mail: keyainternationalpressllp@gmail.com

Published in India, 2024

ISBN : 9789384568238 (5 Vols. Set)

Typeset by Thomson Press

Printed and bound in India by Thomson Press Ltd.,
New Delhi

Contents

Preface

A gene is a molecular unit of heredity of a living organism. It is a name given to some stretches of DNA and RNA that code for a polypeptide or for an RNA chain that has a function in the organism. Living beings depend on genes, as they specify all proteins and functional RNA chains. Genes hold the information to build and maintain an organism's cells and pass genetic traits to offspring. All organisms have many genes corresponding to various biological traits, some of which are immediately visible, such as eye colour or number of limbs, and some of which are not, such as blood type, increased risk for specific diseases, or the thousands of basic biochemical processes that comprise life.

A modern working definition of a gene is "*a locatable region of genomic sequence, corresponding to a unit of inheritance, which is associated with regulatory regions, transcribed regions, and or other functional sequence regions* ".Colloquial usage of the term *gene* (e.g. "good genes", "hair colour gene") may actually refer to an allele: a *gene* is the basic instruction—a sequence of nucleic acids (DNA or, in the case of certain viruses RNA), while an *allele* is one variant of that gene. Referring to having a gene for a trait is no longer the scientifically accepted usage. In most cases, all people would have a gene for the trait in question, but certain people will have a specific allele of that gene, which results in the trait variant. Further, genes code for proteins, which might result in identifiable traits, but it is the gene, not the trait, which is inherited. When proteins are manufactured, the gene is first copied into RNA as an intermediate product. In other cases, the RNA molecules are the actual functional products. For example, RNAs known as ribozymes are capable of enzymatic function, and microRNA has a regulatory role. The DNA sequences from which such RNAs are transcribed are known as RNA genes. Some viruses store their entire genomes in the form of RNA, and contain no DNA at all. Because they use RNA to store genes, their cellular hosts may synthesize their proteins as soon as they are infected and without the delay in waiting for transcription. On the other hand, RNA retroviruses, such as HIV, require the reverse transcription of their genome from RNA into DNA before their proteins can be synthesized. In 2006, French researchers came across a puzzling example of RNA-mediated inheritance in mice. Mice with a loss-of-function mutation in the gene Kit have white tails. Offspring of these mutants can have white tails despite having only normal Kit genes. The research team traced this effect back to mutated Kit RNA. While RNA is common as genetic storage material in viruses, in mammals in particular RNA inheritance has been observed very rarely. The vast majority of living organisms encode their genes in long strands of DNA (deoxyribonucleic acid). DNA consists of a chain made from four types of nucleotide subunits, each composed of: a five-carbon sugar, a phosphate group, and one of the four bases adenine, cytosine, guanine, and thymine. The most common form of DNA in a cell is in a double helix structure, in which two individual DNA strands twist around each other in a right-handed spiral. In this structure, the base pairing rules specify that guanine pairs with cytosine and adenine pairs with thymine. The base pairing between guanine and cytosine forms three hydrogen bonds, whereas the base pairing between adenine and thymine forms two hydrogen bonds. The two strands in a double helix must therefore be *complementary*, that is, their bases must align such that the adenines of one strand are paired with the thymines of the other strand, and so on. The notion of a gene is evolving with the science of genetics, which began when Gregor Mendel noticed that biological variations are inherited from parent organisms as specific, discrete traits. The biological entity responsible for defining traits was later termed a *gene*, but the biological basis for inheritance remained unknown until DNA was identified as the genetic material in the 1940s. Prior to Mendel's work, the dominant theory of heredity was one of blending inheritance, which proposes that the traits of the parents blend or mix in a smooth, continuous gradient in the offspring. Although Mendel's work was largely unrecognized after its first publication in 1866, it was rediscovered in 1900 by three European scientists, Hugo de Vries, Carl Correns, and Erich von Tschermak, who had reached similar conclusions from their own research. However, these scientists were not yet aware of the identity of the 'discrete units' on which genetic material resides.

The overarching theme of the book, the important link between genes, is mirrored in the text. Efforts have been made to present the principles of gene manipulation and its associated techniques, in sufficient detail to enable non-specialist reader to understand them. This book will be useful for students, teachers and researchers.

—Editor

1

Gene: Its Expression

Gene expression is the process by which information from a gene is used in the synthesis of a functional gene product. These products are often proteins, but in non-protein coding genes such as ribosomal RNA (rRNA) genes or transfer RNA (tRNA) genes, the product is a functional RNA. The process of gene expression is used by all known life - eukaryotes (including multicellular organisms), prokaryotes (bacteria and archaea) and viruses - to generate the macromolecular machinery for life. Several steps in the gene expression process may be modulated, including the transcription, RNA splicing, translation, and post-translational modification of a protein. Gene regulation gives the cell control over structure and function, and is the basis for cellular differentiation, morphogenesis and the versatility and adaptability of any organism. Gene regulation may also serve as a substrate for evolutionary change, since control of the timing, location, and amount of gene expression can have a profound effect on the functions (actions) of the gene in a cell or in a multicellular organism.

In genetics, gene expression is the most fundamental level at which the genotype gives rise to the phenotype. The genetic code stored in DNA is "interpreted" by gene expression, and the properties of the expression give rise to the organism's phenotype.

Mechanism

Transcription

The gene itself is typically a long stretch of DNA which carries genetic information encoded by genetic code. Every molecule of DNA consists of two strands, each of them having 5' and 3' ends oriented in anti-parallel direction.

The coding strand contains the genetic information while template strand (non-coding strand) serves as a blueprint for the production of RNA. The production of RNA copies of the DNA is called transcription, and is performed by RNA polymerase, which adds one RNA nucleotide at a time to a growing RNA strand. This RNA is complementary to the template 3' → 5' DNA strand, which is itself complementary to the coding 5' → 3' DNA strand. Therefore, the resulting 5' → 3' RNA strand is identical to the coding DNA strand with the exception that thymines (T) are replaced with uracils (U) in the RNA. A coding DNA strand reading "ATG" is transcribed as "AUG" in RNA.

Transcription in prokaryotes is carried out by a single type of RNA polymerase, which needs DNA sequence called Pribnow box and sigma factor (σ factor) to start transcription. In eukaryotes, the transcription is done by three types of RNA polymerases, each of them needs special DNA sequence called promoter and a set of DNA-binding proteins - transcription factors to initiate the process. RNA polymerase I is responsible for transcription of rRNA genes, while RNA polymerase II transcribes all protein-coding genes but also some non-coding RNAs (e.g. snRNAs, snoRNAs or long non-coding RNAs) as well. It contains special part called C-terminal domain (CTD) that is rich of serines, which after being phosphorylated accumulate factors necessary for RNA modification and maturation. RNA polymerase III transcribes 5S rRNA and tRNA genes but also some small non-coding RNA genes (e.g. 7SK). Transcription ends on a special sequence called terminator.

RNA Processing

While transcription of prokaryotic protein-coding genes creates messenger RNA (mRNA) which is ready for translation, transcription of eukaryotic genes leaves a primary transcript of RNA (pre-mRNA), which first has to undergo series of modification to become a mature mRNA. These include 5' *capping*, which is set of enzymatic reactions that add 7-methylguanosine (m^7G) to the 5' end of pre-mRNA and thus protect the RNA from degradation by exonucleases. The m^7G cap is then bound by cap binding complex heterodimer (CBC20/CBC80) which aids in mRNA export to cytoplasm and also protect the RNA from decapping.

Another modification is 3' *clevage and polyadenylation*. They occur if polyadenylation signal sequence (5'- AAUAAA-3') is present in pre-mRNA,which is usually between protein-coding sequence and terminator. The pre-mRNA is first cleaved and then a series of ~200 adenines (A) are added to form poly(A) tail which protects the RNA from degradation. Poly(A) tail is bound by multiple poly(A)-binding proteins (PABP) necessary for mRNA export and translation re-iniciation. Very important modification of eukaryotic pre-mRNA is *RNA splicing*. Majority of eukaryotic pre-mRNAs consist of alternating segments called exons and introns. During the process of splicing, RNA-protein catalytical complex known as spliceosome, catalyze two transesterification reactions, which remove intron and release it in form of lariat structure and then splice neighbouring exons together. In certain cases, some introns or exons can be either removed or retained in mature mRNA. This so-called alternative splicing creates series of different transcripts originating from a single gene. Because these transcripts can be potentially translated into different proteins, splicing extends the complexity of eukaryotic gene expression.

Extensive RNA processing may be an evolutionary advantage made possible by the nucleus of eukaryotes. In prokaryotes transcription and translation happen together whilst in eukaryotes the nuclear membrane separates the two processes giving time for RNA processing to occur.

Non-coding RNA Maturation

In most organisms non-coding genes (ncRNA) are transcribed as precursors which undergo further processing. In the case of ribosomal RNAs (rRNA), they are often transcribed as a pre-rRNA which contains one or more rRNAs, the pre-rRNA is cleaved and modified (22 -O-methylation and pseudouridine formation) at a specific sites by approximately 150 different small nucleolus-restricted RNA species, called snoRNAs.

SnoRNAs associate with proteins, forming snoRNPs. While snoRNA part basepair with the target RNA and thus position the modification to precise site, the protein part performs the catalytical reaction. In eukaryotes, in particular a snoRNP, called RNase MRP cleaves the 45S pre-rRNA into the 28S, 5.8S, and 18S rRNAs. The rRNA and RNA processing factors form large aggregates called the nucleolus.

In the case of transfer RNA (tRNA), for example, the 5' sequence is removed by RNase P, whereas the 3' end is removed by the tRNase Z enzyme and the non-templated 3' CCA tail is added by a nucleotidyl transferase. In the case of micro RNA (miRNA), miRNAs are first transcribed as primary transcripts or pri-miRNA with a cap and poly-A tail and processed to short, 70-nucleotide stem-loop structures known as pre-miRNA in the cell nucleus by the enzymes Drosha and Pasha.

After being exported, it is then processed to mature miRNAs in the cytoplasm by interaction with the endonuclease Dicer, which also initiates the formation of the RNA-induced silencing complex (RISC), composed of the Argonaute protein.

Even snRNAs and snoRNAs themselves undergo series of modification before they become part of functional RNP complex. This is done either in the nucleoplasm or in the specialized compartments called Cajal bodies. Their bases are methylated or pseudouridinilated by a group of small Cajal body-specific RNAs (scaRNAs) which are structurally similar to snoRNAs.

RNA Export

In eukaryotes most mature RNA must be exported to the cytoplasm from the nucleus. While some RNAs function in the nucleus, many RNAs are transported through the nuclear pores and into the cytosol. Notably this includes all RNA types involved in protein synthesis. In some cases RNAs are additionally transported to a specific part of the cytoplasm, such as a synapse; they are then towed by motor proteins that bind through linker proteins to specific sequences (called "zipcodes") on the RNA.

Translation

For some RNA (non-coding RNA) the mature RNA is the final gene product. In the case of messenger RNA (mRNA) the RNA is an information carrier coding for the synthesis of one or more proteins. mRNA carrying a single protein sequence (common in eukaryotes) is monocistronic whilst mRNA carrying multiple protein sequences (common in prokaryotes) is known as polycistronic. Every mRNA consists of three parts - 5' untranslated region (5'UTR), protein-coding region or open reading frame (ORF) and 3' untranslated region (3'UTR). Coding region carries information for protein synthesis encoded by genetic code into form of triplets. Each triplet of nucleotides of the coding region is called codon and corresponds to a binding site complementary to an anticodon triplet in transfer RNA. Transfer RNAs with the same anticodon sequence always carry identical type of amino acid. Amino acids are then chained together by the ribosome according to order of triplets in the coding region. The ribosome helps transfer RNA to bind to messenger RNA and takes the amino acid from each transfer RNA and makes a structure-less protein out of it.

In prokaryotes translation generally occurs at the point of transcription (co-transcriptionally), often using a messenger RNA which is still in the process of being created. In eukaryotes translation can occur in a variety of regions of the cell depending on where the protein being written is supposed to be. Major locations are the cytoplasm for soluble cytoplasmic proteins and the membrane of endoplasmic reticulum for proteins which are for export from the cell or insertion into a cell membrane. Proteins which are supposed to be expressed at the endoplasmic reticulum are recognised part-way through the translation process. This is governed by the signal recognition particle - a protein which binds to the ribosome and directs it to the endoplasmic reticulum when it finds a signal sequence on the growing (nascent) amino acid chain.

Folding

The polypeptide folds into its characteristic and functional three-dimensional structure from random coil. Each protein exists as an unfolded polypeptide or random coil when translated from a sequence of mRNA to a linear chain of amino acids. This polypeptide lacks any developed three-dimensional structure. Amino acids interact with each other to produce a well-defined three-dimensional structure, the folded protein, known as the native state. The resulting three-dimensional structure is determined by the amino acid sequence (Anfinsen's dogma).

The correct three-dimensional structure is essential to function, although some parts of functional proteins may remain unfolded Failure to fold into the intended shape usually produces inactive proteins with different properties including toxic prions. Several neurodegenerative and other diseases are believed to result from the accumulation of *misfolded* (incorrectly folded) proteins. Many allergies are caused by the folding of the proteins, for the immune system does not produce antibodies for certain protein structures. Enzymes called chaperones assist the newly formed protein to attain (fold into) the 3-dimensional structure it needs to function. Similarly, RNA chaperones help RNAs attain their functional shapes. Assisting protein folding is one of the main roles of the endoplasmic reticulum in eukaryotes.

Protein Transport

Many proteins are destined for other parts of the cell than the cytosol and a wide range of signalling sequences are used to direct proteins to where they are supposed to be. In prokaryotes this is normally a simple process due to limited compartmentalisation of the cell. However in eukaryotes there is a great variety of different targeting processes to ensure the protein arrives at the correct organelle. Not all proteins remain within the cell and many are exported, for example digestive enzymes, hormones and extracellular matrix proteins. In eukaryotes the export pathway is well developed and the main mechanism for the export of these proteins is translocation to the endoplasmic reticulum, followed by transport via the Golgi apparatus.

Regulation of Gene Expression

Regulation of gene expression refers to the control of the amount and timing of appearance of the functional product of a gene. Control of expression is vital to allow a cell to produce the gene products it needs when it needs them; in turn this gives cells the flexibility to adapt to a variable environment, external signals, damage to the cell, etc. Some simple examples of where gene expression is important are:

- Control of Insulin expression so it gives a signal for blood glucose regulation
- X chromosome inactivation in female mammals to prevent an "overdose" of the genes it contains.
- Cyclin expression levels control progression through the eukaryotic cell cycle.

More generally gene regulation gives the cell control over all structure and function, and is the basis for cellular differentiation, morphogenesis and the versatility and adaptability of any organism. Any step of gene expression may be modulated, from the DNA-RNA transcription step to post-translational modification of a protein. The stability of the final gene product, whether it is RNA or protein, also contributes to the expression level of the gene - an unstable product results in a low expression level. In general gene expression is regulated through changes in the number and type of interactions between molecules that collectively influence transcription of DNA and translation of RNA.

Numerous terms are used to describe types of genes depending on how they are regulated, these include:

- A *constitutive gene* is a gene that is transcribed continually compared to a facultative gene which is only transcribed when needed.
- A *housekeeping gene* is typically a constitutive gene that is transcribed at a relatively constant level. The housekeeping gene's products are typically needed for maintenance of the cell. It is generally assumed that their expression is unaffected by experimental conditions. Examples include actin, GAPDH and ubiquitin.
- A *facultative gene* is a gene which is only transcribed when needed compared to a constitutive gene.
- An *inducible gene* is a gene whose expression is either responsive to environmental change or dependent on the position in the cell cycle.

Transcriptional Regulation

Regulation of transcription can be broken down into three main routes of influence; genetic (direct interaction of a control factor with the gene), modulation (interaction of a control factor with the transcription machinery) and epigenetic (non-sequence changes in DNA structure which influence transcription). Direct interaction with DNA is the simplest and the most direct method by which a protein can change transcription levels. Genes often have several protein binding sites around the coding region with the specific function of regulating transcription.

There are many classes of regulatory DNA binding sites known as enhancers, insulators, repressors and silencers. The mechanisms for regulating transcription are very varied, from blocking key binding sites on the DNA for RNA polymerase to acting as an activator and promoting transcription by assisting RNA polymerase binding. The activity of transcription factors is further modulated by intracellular signals causing protein post-translational modification including phosphorylated, acetylated, or glycosylated. These changes influence a transcription factor's ability to bind, directly or indirectly, to promoter DNA, to recruit RNA polymerase, or to favor elongation of a newly synthetized RNA molecule.

The nuclear membrane in eukaryotes allows further regulation of transcription factors by the duration of their presence in the nucleus which is regulated by reversible changes in their structure and by binding of other proteins. Environmental stimuli or endocrine signals may cause modification of regulatory proteins eliciting cascades of intracellular signals, which result in regulation of gene expression.

More recently it has become apparent that there is a huge influence of non-DNA-sequence specific effects on translation. These effects are referred to as epigenetic and involve the higher order structure of DNA, non-sequence specific DNA binding proteins and chemical modification of DNA.

In general epigenetic effects alter the accessibility of DNA to proteins and so modulate transcription. DNA methylation is a widespread mechanism for epigenetic influence on gene expression and is seen in bacteria and eukaryotes and has roles in heritable transcription silencing and transcription regulation. In eukaryotes the structure of chromatin, controlled by the histone code, regulates access to DNA with significant impacts on the expression of genes in euchromatin and heterochromatin areas.

Post-transcriptional Regulation

In eukaryotes, where export of RNA is required before translation is possible, nuclear export is thought to provide additional control over gene expression. All transport in and out of the nucleus is via the nuclear pore and transport is controlled by a wide range of importin and exportin proteins. Expression of a gene coding for a protein is only possible if the messenger RNA carrying the code survives long enough to be translated.

In a typical cell an RNA molecule is only stable if specifically protected from degradation. RNA degradation has particular importance in regulation of expression in eukaryotic cells where mRNA has to travel significant distances before being translated.

In eukaryotes RNA is stabilised by certain post-transcriptional modifications, particularly the 5' cap and poly-adenylated tail. Intentional degradation of mRNA is used not just as a defence mechanism from foreign RNA (normally from viruses) but also as a route of mRNA *destabilisation*. If an mRNA molecule has a complementary sequence to a small interfering RNA then it is targeted for destruction via the RNA interference pathway.

Translational Regulation

Direct regulation of translation is less prevalent than control of transcription or mRNA stability but is occasionally used. Inhibition of protein translation is a major target for toxins and antibiotics in order to kill a cell by overriding its normal gene expression control. Protein synthesis inhibitors include the antibiotic neomycin and the toxin ricin.

Protein Degradation

Once protein synthesis is complete the level of expression of that protein can be reduced by protein degradation. There are major protein degradation pathways in all prokaryotes and eukaryotes of which the proteasome is a common component. An unneeded or damaged protein is often labelled for degradation by addition of ubiquitin.

Measurement

Measuring gene expression is an important part of many life sciences - the ability to quantify the level at which a particular gene is expressed within a cell, tissue or organism can give a huge amount of information. For example measuring gene expression can:

- Identify viral infection of a cell (viral protein expression)
- Determine an individual's susceptibility to cancer (oncogene expression)
- Find if a bacterium is resistant to penicillin (beta-lactamase expression).

Similarly the analysis of the location of expression protein is a powerful tool and this can be done on an organism or cellular scale. Investigation of localisation is particularly important for study of development in multicellular organisms and as an indicator of protein function in single cells. Ideally measurement of expression is done by detecting the final gene product (for many genes this is the protein) however it is often easier to detect one of the precursors, typically mRNA, and infer gene expression level.

mRNA Quantification

Levels of mRNA can be quantitatively measured by Northern blotting which gives size and sequence information about the mRNA molecules. A sample of RNA is separated on an agarose gel and hybridized to a radio-labeled RNA probe that is complementary to the target sequence. The radio-labeled RNA is then detected by an autoradiograph. The main problems with Northern blotting stem from the use of radioactive reagents (which make the procedure time consuming and potentially dangerous) and lower quality quantification than more modern methods (due to the fact that quantification is done by measuring band strength in an image of a gel). Northern blotting is, however, still widely used as the additional mRNA size information allows the discrimination of alternately spliced transcripts.

A more modern low-throughput approach for measuring mRNA abundance is reverse transcription quantitative polymerase chain reaction (RT-PCR followed with qPCR). RT-PCR first generates a DNA template from the mRNA by reverse transcription, which is called cDNA. This cDNA template is then used for qPCR where the change in fluorescence of a probe changes as the DNA amplification process progresses. With a carefully constructed standard curve qPCR can produce an absolute measurement such as number of copies of mRNA, typically in units of copies per nanolitre of homogenized tissue or copies per cell. qPCR is very sensitive (detection of a single mRNA molecule is possible), but can be expensive due to the fluorescent probes required.

An even more advanced approach is to individually tag single mRNA molecules with fluorescent barcodes (nanostrings), which can be detected one-by-one and counted for direct digital quantification. The advantage of this approach is that it does not rely on analog quantification of fluorescent intensity, which can be problematic due to noise, lack of linearity, and narrow dynamic range. Instead, the technique relies on fluorescence to detect simply the presence of a single mRNA molecule in a binary ("yes" or "no") mode. This method was invented by Dr. Krassen Dimitrov at the Institute for Systems Biology and commercialized through his start-up company, NanoString Technologies

Northern blots and RT-qPCR are good for detecting whether a single gene is being expressed, but it quickly becomes impractical if many genes within the sample are being studied. Using DNA microarrays transcript levels for many genes at once (expression profiling) can be measured. Recent advances in microarray technology allow for the quantification, on a single array, of transcript levels for every known gene in several organism's genomes, including humans. Alternatively "tag based" technologies like Serial analysis of gene expression (SAGE), which can provide a relative measure of the cellular concentration of different mRNAs, can be used. The great advantage of tag-based methods is the "open architecture", allowing for the exact measurement of any transcript, with a known or unknown sequence.

Protein Quantification

For genes encoding proteins the expression level can be directly assessed by a number of means with some clear analogies to the techniques for mRNA quantification. The most commonly used method is to perform a Western blot against the protein of interest - this gives information on the size of the protein in addition to its identity. A sample (often cellular lysate) is separated on a polyacrylamide gel, transferred to a membrane and then probed with an antibody to the protein of interest. The antibody can either be conjugated to a fluorophore or to horseradish peroxidase for imaging and/or quantification. The gel-based nature of this assay makes quantification less accurate but it has the advantage of being able to identify later modifications to the protein, for example proteolysis or ubiquitination, from changes in size.

Localisation

In situ-hybridization of Drosophila embryos at different developmental stages for the mRNA responsible for the expression of hunchback. High intensity of blue color marks places with high hunchback mRNA quantity. Analysis of expression is not limited to only quantification; localisation can also be determined. mRNA can be detected with a suitably labelled complementary mRNA strand and protein can be detected via labelled antibodies. The probed sample is then observed by microscopy to identify where the mRNA or protein is. The three-dimensional structure of green fluorescent protein. The residues in the centre of the "barrel" are responsible for production of green light after exposing to higher energetic blue light. From PDB 1EMA.

By replacing the gene with a new version fused a green fluorescent protein (or similar) marker expression may be directly quantified in live cells. This is done by imaging using a fluorescence microscope. It is very difficult to clone a GFP-fused protein into its native location in the genome without affecting expression levels so this method often cannot be used to measure endogenous gene expression. It is, however, widely used to measure the expression of a gene artificially introduced into the cell, for example via an expression vector. It is important to note that by fusing a target protein to a fluorescent reporter the protein's behavior, including its cellular localization and expression level, can be significantly changed.

The enzyme-linked immunosorbent assay works by using antibodies immobilised on a microtiter plate to capture proteins of interest from samples added to the well. Using a detection antibody conjugated to an enzyme or fluorophore the quantity of bound protein can be accurately measured by fluorometric or colourimetric detection. The detection process is very similar to that of a Western blot, but by avoiding the gel steps more accurate quantification can be achieved.

Expression System

An expression system is a system specifically designed for the production of a gene product of choice. This is normally a protein although may also be RNA, such as tRNA or a ribozyme. An expression system consists of a gene, normally encoded by DNA, and the molecular machinery required to transcribe the DNA into mRNA and translate the mRNA into protein using the reagents provided. In the broadest sense this includes every living cell but the term is more normally used to refer to expression as a laboratory tool. An expression system is therefore often artificial in some manner. Expression systems are, however, a fundamentally natural process. Viruses are an excellent example where they replicate by using the host cell as an expression system for the viral proteins and genome.

In Nature

In addition to these biological tools, certain naturally observed configurations of DNA (genes, promoters, enhancers, repressors) and the associated machinery itself are referred to as an expression system. This term is normally used in the case where a gene or set of genes is switched on under well defined conditions. For example the simple repressor switch expression system in Lambda phage and the lac operator system in bacteria. Several natural expression systems are directly used or modified and used for artificial expression systems such as the Tet-on and Tet-off expression system.

Gene Networks

Genes have sometimes been regarded as nodes in a network, with inputs being proteins such as transcription factors, and outputs being the level of gene expression. The node itself performs a function, and the operation of these functions have been interpreted as performing a kind of information processing within cell and determine cellular behavior.

Gene networks can also be constructed without formulating an explicit causal model. This is often the case when assembling networks from large expression data sets. Covariation and correlation of expression is computed across a large sample of cases and measurements (often transcriptome or proteome data). The source of variation can be either experimental or natural (observational). There are several ways to construct gene expression networks, but one common approach is to compute a matrix of all pair-wise correlations of expression across conditions, time points, or individuals and convert the matrix (after thresholding at some cut-off value) into a graphical representation in which nodes represent genes, transcripts, or proteins and edges connecting these nodes represent the strength of association.

Techniques and Tools

The following experimental techniques are used to measure gene expression and are listed in roughly chronological order, starting with the older, more established technologies. They are divided into two groups based on their degree of multiplexity.

- Low-to-mid-plex techniques:
  - Reporter gene
  - Northern blot
  - Western blot
  - Fluorescent in situ hybridization
  - Reverse transcription PCR
  - Digital counting of single transcript molecules.

- Higher-plex techniques:
  - o SAGE
  - o DNA microarray
  - o Tiling array
  - o RNA-Seq.

Transcriptional Bursting

Transcriptional bursting, also known as transcriptional pulsing, is a fundamental property of genes from bacteria to humans. Transcription of genes, the process which transforms the stable code written in DNA into the mobile RNA message can occur in "bursts" or "pulses". This phenomenon has recently come to light with the advent of new technologies, such as MS2 tagging, to detect RNA production in single cells, allowing precise measurements of RNA number, or RNA appearance at the gene. Other, more widespread techniques, such as Northern Blotting, Microarrays, RT-PCR and RNA-Seq, measure bulk RNA levels from homogenous population extracts. These techniques lose dynamic information from individual cells, and give the impression transcription is a continuous smooth process.

The reality is that transcription is irregular, with strong periods of activity, interspersed by long periods of inactivity. Averaged over millions of cells, this appears continuous. But at the individual cell level, there is considerable variability, and for most genes, very little activity at any one time.

Bursting may result from the stochastic nature of biochemical events superimposed upon a 2 or more step fluctuation. In its most simple form, the gene can exist in 2 states, one where activity is negligible and one where there is a certain probability of activation. Only in the second state does transcription readily occur. Whilst the nuclear and signaling landscapes of complex eukaryotic nuclei are likely to favour more than two simple states-

for example, there are over twenty post-translational modifications of nucleosomes known, this simple two step model perhaps provides a reasonable intellectual framework for understanding the changing probabilities affecting transcription. It seems likely that some rudimentary eukaryotes have genes which do not show bursting. The genes are always in the permissive state, with a simple probability describing the numbers of RNAs generated.

What do the repressive and permissive states represent? An attractive idea is that the repressed state is a closed chromatin conformation whilst the permissive state is an open one. Another hypothesis is that the fluctuations reflect transition between bound pre-initiation complexes (permissive) and dissociated ones (restrictive). Bursts may also result from bursty signalling, cell cycle effects or movement of chromatin to and from transcription factories.

The bursting phenomenon, as opposed to simple probabilistic models of transcription, can account for the high variability (transcriptional noise) in gene expression occurring between cells in isogenic populations. This variability in turn can have tremendous consequences on cell behaviour, and must be mitigated or integrated. In certain contexts, such as the survival of microbes in rapidly changing stressful environments, or several types of scattered differentiation, the variability may be essential. Variability also impacts upon the effectiveness of clinical treatment, with resistance of bacteria to antibiotics demonstrably caused by non-genetic differences. Variability in gene expression may also contribute to resistance of sub-populations of cancer cells to chemotherapy.

Expression Profiling

In the field of molecular biology, gene expression profiling is the measurement of the activity (the expression) of thousands of genes at once, to create a global picture of cellular function. These profiles can, for example, distinguish between cells that are actively dividing, or show how the cells react to a particular treatment. Many experiments of this sort measure an entire genome simultaneously, that is, every gene present in a particular cell.

DNA Microarray technology measures the relative activity of previously identified target genes. Sequence based techniques, like serial analysis of gene expression (SAGE, SuperSAGE) are also used for gene expression profiling. SuperSAGE is especially accurate and can measure any active gene, not just a predefined set. The advent of next-generation sequencing has made sequence based expression analysis an increasingly popular, "digital" alternative to microarrays. However, microarrays are far more common, accounting for 17,000 PubMed articles by 2006.

Background

Expression profiling is a logical next step after sequencing a genome: the sequence tells us what the cell could possibly do, while the expression profile tells us what it is actually doing now. Genes contain the instructions for making messenger RNA (mRNA), but at any moment each cell makes mRNA from only a fraction of the genes it carries. If a gene is used to produce mRNA, it is considered "on", otherwise "off".

Many factors determine whether a gene is on or off, such as the time of day, whether or not the cell is actively dividing, its local environment, and chemical signals from other cells. Skin cells, liver cells and nerve cells turn on (express) somewhat different genes and that is in large part what makes them different. Therefore, an expression profile allows one to deduce a cell's type, state, environment, and so forth. Expression profiling experiments often involve measuring the relative amount of mRNA expressed in two or more experimental conditions. This is because altered levels of a specific sequence of mRNA suggest a changed need for the protein coded for by the mRNA, perhaps indicating a homeostatic response or a pathological condition. For example, higher levels of mRNA coding for alcohol dehydrogenase suggest that the cells or tissues under study are responding to increased levels of ethanol in their environment. Similarly, if breast cancer cells express higher levels of mRNA associated with a particular transmembrane receptor than normal cells do, it might be that this receptor plays a role in breast cancer. A drug that interferes with this receptor may prevent or treat breast cancer.

In developing a drug, one may perform gene expression profiling experiments to help assess the drug's toxicity, perhaps by looking for changing levels in the expression of cytochrome P450 genes, which may be a biomarker of drug metabolism. Gene expression profiling may become an important diagnostic test.

Comparison to Proteomics

The human genome contains on the order of 25,000 genes which work in concert to produce on the order of 1,000,000 distinct proteins. This is due to alternative splicing, and also because cells make important changes to proteins through posttranslational modification after they first construct them, so a given gene serves as the basis for many possible versions of a particular protein. In any case, a single mass spectrometry experiment can identify about 2,000 proteins or 0.2% of the total. While knowledge of the precise proteins a cell makes (proteomics) is more relevant than knowing how much messenger RNA is made from each gene, gene expression profiling provides the most global picture possible in a single experiment.

Use in Hypothesis Generation and Testing

Sometimes, a scientist already has an idea what is going on, a hypothesis, and he or she performs an expression profiling experiment with the idea of potentially disproving this hypothesis. In other words, the scientist is making a specific prediction about levels of expression that could turn out to be false.

More commonly, expression profiling takes place before enough is known about how genes interact with experimental conditions for a testable hypothesis to exist. With no hypothesis, there is nothing to disprove, but expression profiling can help to identify a candidate hypothesis for future experiments.

Most early expression profiling experiments, and many current ones, have this form which is known as class discovery. A popular approach to class discovery involves grouping similar genes or samples together using k-means or hierarchical clustering. The figure above represents the output of a two dimensional cluster, in which similar samples (rows, above) and similar gene probes (columns) were organized so that they would lie close together. The simplest form of class discovery would be to list all the genes that changed by more than a certain amount between two experimental conditions. Class prediction is more difficult than class discovery, but it allows one to answer questions of direct clinical significance such as, given this profile, what is the probability that this patient will respond to this drug? This requires many examples of profiles that responded and did not respond, as well as cross-validation techniques to discriminate between them.

Limitations

In general, expression profiling studies report those genes that showed statistically significant differences under changed experimental conditions. This is typically a small fraction of the genome for several reasons.

First, different cells and tissues express a subset of genes as a direct consequence of cellular differentiation so many genes are turned off.

Second, many of the genes code for proteins that are required for survival in very specific amounts so many genes do not change.

Third, cells use many other mechanisms to regulate proteins in addition to altering the amount of mRNA, so these genes may stay consistently expressed even when protein concentrations are rising and falling.

Fourth, financial constraints limit expression profiling experiments to a small number of observations of the same gene under identical conditions, reducing the statistical power of the experiment, making it impossible for the experiment to identify important but subtle changes.

Finally, it takes a great amount of effort to discuss the biological significance of each regulated gene, so scientists often limit their discussion to a subset. Newer microarray analysis techniques automate certain aspects of attaching biological significance to expression profiling results, but this remains a very difficult problem.

The relatively short length of gene lists published from expression profiling experiments limits the extent to which experiments performed in different laboratories appear to agree. Placing expression profiling results in a publicly accessible microarray database makes it possible for researchers to assess expression patterns beyond the scope of published results, perhaps identifying similarity with their own work.

Validation of High Throughput Measurements

Both DNA microarrays and qPCR exploit the preferential binding or "base pairing" of complementary nucleic acid sequences, and both are used in gene expression profiling, often in a serial fashion. While high throughput DNA microarrays lack the quantitative accuracy of qPCR, it takes about the same time to measure the gene expression of a few dozen genes via qPCR as it would to measure an entire genome using DNA microarrays. So it often makes sense to perform semi-quantitative DNA microarray analysis experiments to identify candidate genes, then perform qPCR on some of the most interesting candidate genes to validate the microarray results. Other experiments, such as a Western blot of some of the protein products of differentially expressed genes, make conclusions based on the expression profile more persuasive, since the mRNA levels do not necessarily correlate to the amount of expressed protein.

Statistical Analysis

Data analysis of microarrays has become an area of intense research. Simply stating that a group of genes were regulated by at least twofold, once a common practice, lacks a solid statistical footing. With five or fewer replicates in each group, typical for microarrays, a single outlier observation can create an apparent difference greater than two-fold. In addition, arbitrarily setting the bar at two-fold is not biologically sound, as it eliminates from consideration many genes with obvious biological significance.

Rather than identify differentially expressed genes using a fold change cutoff, one can use a variety of statistical tests or omnibus tests such as ANOVA, all of which consider both fold change and variability to create a p-value, an estimate of how often we would observe the data by chance alone. Applying p-values to microarrays is complicated by the large number of multiple comparisons (genes) involved.

For example, a p-value of .05 is typically thought to indicate significance, since it estimates a 5% probability of observing the data by chance. But with 10,000 genes on a microarray, 500 genes would be identified as significant at $p < .05$ even if there were no difference between the experimental groups.

One obvious solution is to consider significant only those genes meeting a much more stringent p value criterion, e.g., one could perform a Bonferroni correction on the p-values, or use a false discovery rate calculation to adjust p-values in proportion to the number of parallel tests involved.

Unfortunately, these approaches may reduce the number of significant genes to zero, even when genes are in fact differentially expressed. Current statistics such as Rank products aim to strike a balance between false discovery of genes due to chance variation and non-discovery of differentially expressed genes. Commonly cited methods include the Significance Analysis of Microarrays (SAM) and a wide variety of methods are available from Bioconductor and a variety of analysis packages from bioinformatics companies.

Selecting a different test usually identifies a different list of significant genes since each test operates under a specific set of assumptions, and places a different emphasis on certain features in the data. Many tests begin with the assumption of a normal distribution in the data, because that seems like a sensible starting point and often produces results that appear more significant. Some tests consider the joint distribution of all gene observations to estimate general variability in measurements, while others look at each gene in isolation. Many modern microarray analysis techniques involve bootstrapping (statistics), machine learning or Monte Carlo methods. As the number of replicate measurements in a microarray experiment increases, various statistical approaches yield increasingly similar results, but lack of concordance between different statistical methods makes array results appear less trustworthy. The MAQC Project makes recommendations to guide researchers in selecting more standard methods so that experiments performed in different laboratories will agree better.

Different from the analysis on differentially expressed individual genes, another type of analysis focuses on differential expression or perturbation of pre-defined gene sets and is called gene set analysis. Gene set analysis demonstrated several major advantages over individual gene differential expression analysis. Gene sets are groups of genes that are functionally related according to current knowledge. Therefore, gene set analysis is considered a knowledge based analysis approach. Commonly used gene sets include those derived from KEGG pathways, Gene Ontology terms, gene groups that share some other functional annotations, such as common transcriptional regulators etc. Representative gene set analysis methods include GSEA, which estimates significance of gene sets based on permutation of sample labels, and GAGE, which tests the significance of gene sets based on permutation of gene labels or a parametric distribution.

Gene Annotation

While the statistics may reliably identify which gene products change under experimental conditions, making biological sense of expression profiling rests on knowing which protein each gene product makes and what function this protein performs.

Gene annotation provides functional and other information, for example the location of each gene within a particular chromosome. Some functional annotations are more reliable than others; some are absent. Gene annotation databases change regularly, and various databases refer to the same protein by different names, reflecting a changing understanding of protein function. Use of standardized gene nomenclature helps address the naming aspect of the problem, but exact matching of transcripts to genes remains an important consideration.

Categorizing Regulated Genes

Having identified some set of regulated genes, the next step in expression profiling involves looking for patterns within the regulated set. Do the proteins made from these genes perform similar functions? Are they chemically similar? Do they reside in similar parts of the cell? Gene ontology analysis provides a standard way to define these relationships. Gene ontologies start with very broad categories, e.g., "metabolic process" and break them down into smaller categories, e.g., "carbohydrate metabolic process" and finally into quite restrictive categories like "inositol and derivative phosphorylation".Genes have other attributes beside biological function, chemical properties and cellular location. One can compose sets of genes based on proximity to other genes, association with a disease, and relationships with drugs or toxins.

The Molecular Signatures Database and the Comparative Toxicogenomics Database are examples of resources to categorize genes in numerous ways.

Finding Patterns among Regulated Genes

Regulated genes are categorized in terms of what they are and what they do, important relationships between genes may emerge. For example, we might see evidence that a certain gene creates a protein to make an enzyme that activates a protein to turn on a second gene on our list. This second gene may be a transcription factor that regulates yet another gene from our list. Observing these links we may begin to suspect that they represent much more than chance associations in the results, and that they are all on our list because of an underlying biological process.

On the other hand, it could be that if one selected genes at random, one might find many that seem to have something in common. In this sense, we need rigorous statistical procedures to test whether the emerging biological themes is significant or not. That's where gene set analysis comes in.

Cause and Effect Relationships

Fairly straightforward statistics provide estimates of whether associations between genes on lists are greater than what one would expect by chance. These statistics are interesting, even if they represent a substantial oversimplification of what is really going on. Here is an example. Suppose there are 10,000 genes in an experiment, only 50 (0.5%) of which play a known role in making cholesterol.

The experiment identifies 200 regulated genes. Of those, 40 (20%) turn out to be on a list of cholesterol genes as well. Based on the overall prevalence of the cholesterol genes (0.5%) one expects an average of 1 cholesterol gene for every 200 regulated genes, that is, 0.005 times 200. This expectation is an average, so one expects to see more than one some of the time. The question becomes how often we would see 40 instead of 1 due to pure chance. According to the hypergeometric distribution one would expect to try about 10^57 times (10 followed by 56 zeroes) before picking 39 or more of the chlolesterol genes from a pool of 10,000 by drawing 200 genes at random. Whether one pays much attention to how infinitesimally small the probability of observing this by chance is, one would conclude that the regulated gene list is enriched in genes with a known cholesterol association.

One might further hypothesize that the experimental treatment regulates cholesterol, because the treatment seems to selectively regulate genes associated with cholesterol. While this may be true, there are a number of reasons why making this a firm conclusion based on enrichment alone represents an unwarranted leap of faith. One previously mentioned issue has to do with the observation that gene regulation may have no direct impact on protein regulation: even if the proteins coded for by these genes do nothing other than make cholesterol, showing that their mRNA is altered does not directly tell us what is happening at the protein level.

It is quite possible that the amount of these cholesterol-related proteins remains constant under the experimental conditions. Second, even if protein levels do change, perhaps there is always enough of them around to make cholesterol as fast as it can be possibly made, that is, another protein, not on our list, is the rate determining step in the process of making cholesterol.

Finally, proteins typically play many roles, so these genes may be regulated not because of their shared association with making cholesterol but because of a shared role in a completely independent process.

Bearing the forgoing caveats in mind, while gene profiles do not in themselves prove causal relationships between treatments and biological effects, they do offer unique biological insights that would often be very difficult to arrive at in other ways.

Using Patterns to Find Regulated Genes

As described above, one can identify significantly regulated genes first and then find patterns by comparing the list of significant genes to sets of genes known to share certain associations. One can also work the problem in reverse order. Here is a very simple example. Suppose there are 40 genes associated with a known process, for example, a predisposition to diabetes. Looking at two groups of expression profiles, one for mice fed a high carbohydrate diet and one for mice fed a low carbohydrate diet, one observes that all 40 diabetes genes are expressed at a higher level in the high carbohydrate group than the low carbohydrate group. Regardless of whether any of these genes would have made it to a list of significantly altered genes, observing all 40 up, and none down appears unlikely to be the result of pure chance: flipping 40 heads in a row is predicted to occur about one time in a trillion attempts using a fair coin.

For a type of cell, the group of genes whose combined expression pattern is uniquely characteristic to a given condition constitutes the gene signature of this condition. Ideally, the gene signature can be used to select a group of patients at a specific state of a disease with accuracy that facilitates selection of treatments. Gene Set Enrichment Analysis (GSEA) and similar methods take advantage of this kind of logic but uses more sophisticated statistics, because component genes in real processes display more complex behavior than simply moving up or down as a group, and the amount the genes move up and down is meaningful, not just the direction. In any case, these statistics measure how different the behavior of some small set of genes is compared to genes not in that small set.

GSEA uses a Kolmogorov Smirnov style statistic to see whether any previously defined gene sets exhibited unusual behavior in the current expression profile. This leads to a multiple hypothesis testing challenge, but reasonable methods exist to address it.

Conclusions

Expression profiling provides new information about what genes do under various conditions. Overall, microarray technology produces reliable expression profiles. From this information one can generate new hypotheses about biology or test existing ones. However, the size and complexity of these experiments often results in a wide variety of possible interpretations. In many cases, analyzing expression profiling results takes far more effort than performing the initial experiments.

Most researchers use multiple statistical methods and exploratory data analysis before publishing their expression profiling results, coordinating their efforts with a biostatistician or other expert in microarray technology. Good experimental design, adequate biological replication and follow up experiments play key roles in successful expression profiling experiments.

Paramutation

In epigenetics, paramutation is an interaction between two alleles of a single locus, resulting in a heritable change of one allele that is induced by the other allele. Paramutation violates Mendel's first law, which states that in the process of the formation of the gametes (egg or sperm) the allelic pairs separate, one going to each gamete, and that each allele remains completely uninfluenced by the other.

In paramutation an allele in one generation heritably affects the other allele in future generations, even if the allele causing the change is itself not transmitted. What may be transmitted are patterns of DNA methylation or RNAs such as piRNAs, siRNAs, miRNAs or other regulatory RNAs. Through proper breeding, paramutation can result in isogenic sibling plants with drastically different phenotypes.

Paramutation was first discovered and studied in maize (*Zea mays*) by R.A. Brink at the University of Wisconsin–Madison in the 1950s. Brink noticed that specific weakly expressed alleles of the *red1 (r1)* locus in maize, which encodes a transcription factor that confers red pigment to corn kernels, can heritably change specific strongly expressed alleles to a weaker expression state.

The weaker expression state adopted by the changed allele is heritable and can, in turn, change the expression state of other active alleles in a process termed secondary paramutation. Brink showed that the influence of the paramutagenic allele could persist for many generations.

Interestingly, paramutation can result in a single allele of a gene controlling a spectrum of phenotypes. At *r1* in maize, for example, the weaker expression state adopted by an allele following paramutation can range from completely colorless to nearly fully-colored kernels. This is an exception to the general observation that continuous variation is controlled by many genes.

Allelic interactions similar to paramutation have since been reported in other organisms, including tomato, pea, and mice.

The molecular basis of paramutation is being unraveled, almost exclusively in maize. Paramutation may share common mechanisms with other epigenetic phenomena, such as gene silencing and genomic imprinting.

In maize, paramutation seems to share many traits with the well understood RNA-directed DNA-methylation pathway in *Arabidopsis thaliana*, even though it has never been observed in the famous model plant.

Alleman (2006) reported that, in maize, "paramutation is RNA-directed. Stability of the chromatin states associated with paramutation and transposon silencing requires the *mop1* gene, which encodes an RNA-dependent RNA polymerase."

Exactly how the RNA produced by this polymerase causes paramutation in maize is not yet understood, but like other epigenetic changes, it involves the covalent modification of DNA and/or the DNA-bound histone proteins without changing the DNA sequence itself.

Ridges

Ridges (regions of increased gene expression) are domains of the genome with a high gene expression; the opposite of ridges are antiridges. The term was first used by Caron et al. in 2001. Characteristics of ridges are:

- Gene dense
- Contain much C and G nucleobases
- Genes have short introns

- high SINE repeat density
- low LINE repeat density.

Discovery

Clustering of genes in prokaryotes was known for a long time. Their genes are grouped in operons, genes within operons share a common promoter unit. These genes are mostly functionally related. The genome of prokaryotes is relatively very simple and compact. In eukaryotes the genome is huge and only a small amount of it are functionally genes, furthermore the genes are not arranged in operons. Except for nematodes and trypanosomes; although their operons are different from the prokaryotic operons. In eukaryotes each gene has a transcription regulation site of its own. Therefore genes don't have to be in close proximity to be co-expressed.

Therefore it was long assumed that eukaryotic genes were randomly distributed across the genome due to the high rate of chromosome rearrangements. But because the complete sequence of genomes became available it became possible to absolutely locate a gene and measure its distance to other genes.

The first eukaryote genome ever sequenced was that of Saccharomyces cerevisiae, or budding yeast, in 1996. Half a year after that Velculescu et al. (1997) published a research in which they had integrated SAGE data with the now available genome map. During a cell cycle different genes are active in a cell. Therefore they used SAGE data from three moments of the cell cycle (log phase, S phase-arrested and G2/M-phase arrested cells). Because in yeast all genes have a promoter unit of their own it was not suspected that genes would cluster near to each other but they did. Clusters were present on all 16 yeast chromosomes. A year later Cho et al. also reported (although in more detail) that certain genes are located near to each other in yeast.

Characteristics and Function

Co-expression

Cho et al. were the first who determined that clustered genes have the same expression levels. They identified transcripts that show cell-cycle dependent periodicity. Of those genes 25% was located in close proximity to other genes which were transcript in the same cell cycle. Cohen et al. (2000) also identified clusters of co-expressed genes.

Caron et al. (2001) made a human transcriptome map of 12 different tissues (cancer cells) and concluded that genes are not randomly distributed across the chromosomes. Instead, genes tent to cluster in groups of sometimes 39 genes in close proximity. Clusters were not only gene dense. They identified 27 clusters of genes with very high expression levels and called them RIDGEs.

A common RIDGE counts 6 to 30 genes per centiray. However, there were great exceptions, 40 to 50% of the RIDGEs were not that gene dense; just like in yeast these RIDGEs were located in the telomere regions. Lercher et al. (2002) pointed to some weaknesses in Caron's approach. Clusters of genes in close proximity and high transcription levels can easily been generated by tandem duplicates. Genes can generate duplicates of themselves which are incorporated in their neighborhood.

These duplicates can either became a functional part of the pathway of their parent gene, or (because they are no longer favored by natural selection) gain deleterious mutations and turn into pseudogenes. Because these duplicates are false positives in the search for gene clusters they have to be excluded. Lercher excluded neighboring genes with high resemblance to each other, after that he searched with a sliding window for regions with 15 neighboring genes.

It was clear that gene dense regions existed. There was a striking correlation between gene density and a high CG content. Some clusters indeed had high expression levels. But most of the highly expressed regions consisted of housekeeping genes; genes that are highly expressed in all tissues because they code for basal mechanisms. Only a minority of the clusters contained genes that were restricted to specific tissues.

Versteeg et al. (2003) tried, with a better human genome map and better SAGE taqs, to determine the characteristics of RIDGEs more specific. Overlapping genes were treated as one gene, and genes without introns were rejected as pseudogenes. They determined that RIDGEs are very gene dense, have a high gene expression, short introns, high SINE repeat density and low LINE repeat density. Clusters containing genes with very low transcription levels had characteristics that were the opposite of RIDGEs, therefore those clusters were called antiridges. LINE repeats are junk DNA which contains a cleavage site of endonuclease (TTTTA).

Their scarcity in RIDGEs can be explained by the fact that natural selection favors the scarcity of LINE repeats in ORFs because their endonuclease sites can cause deleterious mutation to the genes. Why SINE repeats are abundant is not yet understood. Versteeg et al. also concluded that, contrary to Lerchers analysis, the transcription levels of many genes in RIDGEs (for example a cluster on chromosome 9) can vary strongly between different tissues. Lee et al. (2003) analysed the trend of gene clustering between different species. They compared Saccharomyces cerevisiae, Homo sapiens, Caenorhabditis elegans, Arabidopsis thaliana and Drosophila melanogaster, and found a degree of clustering, as fraction of genes in loose clusters, of respectively (37%), (50%), (74%), (52%) and (68%). They concluded that pathways of which the genes are clusters across many species are rare. They found seven universally clustered pathways: glycolysis, aminoacyl-tRNA biosynthesis, ATP synthase, DNA polymerase, hexachlorocyclohexane degradation, cyanoamino acid metabolism, and photosynthesis (ATP synthesis in non plant species). Not surprisingly these are basic cellular pathways. Lee et al. used very diverse groups of animals. Within these groups clustering is conserved, for example the clustering motifs of Homo sapiens and Mus musculus are more or less the same. Spellman and Rubin (2002) made a transcriptome map of Drosophila. Of all assayed genes 20% was clustered. Clusters consisted of 10 to 30 genes over a group size of about 100 kilobases. The members of the clusters were not functionally related and the location of clusters didn't correlate with know chromatin structures.

This study also showed that within clusters the expression levels of on average 15 genes was much the same across the many experimental conditions which were used. These similarities were so striking that the authors reasoned that the genes in the clusters are not individually regulated by their personal promoter but that changes in the chromatin structure were involved. A similar co-regulation pattern was published in the same year by Roy et al. (2002) in C. elegans.

Many genes which are grouped into clusters show the same expression profiles in human invasive ductal breast carcinomas. Roughly 20 % of the genes show a correlation with their neighbors. Clusters of co-expressed genes were divided by regions with less correlation between genes. These clusters could cover an entire chromosome arm. Contrary to previous discussed reports Johnidis et al. (2005) have discovered that (at least some) genes within clusters are not co-regulated. Aire is a transcription factor which has a up- and down-regulation effect on various genes. It functions in negative selection of thymocytes, which responds to the organisms own epitopes, by medullary cells.

The genes that were controlled by aire clustered. 53 of the genes most activated by aire had an aire-activated neighbor within 200 Kb or less, and 32 of the genes most repressed by aire had an aire-repressed neighbor within 200 Kb; this is less than expected by change. They did the same screening for the transcriptional regulator CIITA. These transcription regulators didn't have the same effect on al genes in the same cluster. Genes that were activated and repressed or unaffected were sometimes present in the same cluster. In this case, it's impossible that aire-regulated genes were clustered because they were all co-regulated.

So it is not very clear if domains are co-regulated or not. A very effective way to test this would be by insert synthetic genes into RIDGEs, antiridges and/or random places in the genome and determine their expression. Those expression levels must be compared to each other. Gierman et al. (2007) were the first who proved co-regulation using this approach. As a insertion construct they used a fluorescing GFP gene driven by the ubiquitously expressed human phosphoglycerate kinase (PGK) promoter. They integrated this construct in 90 different positions in the genome of human HEK293 cells. They found that the expression of the construct in Ridges was indeed higher than those inserted in antiridges (while all constructs have the same promoter). They investigated if these differences in expressions were due to genes in the direct neighborhood of the constructs or by the domain as a whole. They found that constructs next to highly expressed genes were slightly more expressed than others. But when to enlarged the window size to the surrounding 49 genes (domain level) they saw that constructs located in domains with an overall high expression had a more than 2-fold higher expression then those located in domains with a low expression level.

They also checked if the construct was expressed at similar levels as neighboring genes, and if that tight co-expression was present solely within RIDGEs. They found that the expressions were highly correlated within RIDGEs, and almost absent near the end and outside the RIDGEs. Previous observations and the research of Gierman et al. proved that the activity of a domain has great impact on the expression of the genes located in it. And the genes within a RIDGE are co-expressed. However the constructs used by Gierman et al. were regulated by al fulltime active promoter.

The genes of the research of Johnidis et al. were dependent of the present of the aire transcription factor. The strange expression of the aire regulated genes could partly have been caused by differences in expression and conformation of the aire transcription factor itself.

Functional Relation

It was known before the genomic era that clustered genes tend to be functionally related. Abderrahim et al. (1994) had shown that all the genes of the major histocompatibility complex were clustered on the 6p21 chromosome. Roy et al. (2002) showed that in the nematode C. elegans genes that are solely expressed in muscle tissue during the larval stage tent to cluster in small groups of 2-5 genes. They identified 13 clusters.

Yamashita et al. (2004) showed that genes related to specific functions in organs tent to cluster. Six liver related domains contained genes for xenobiotic, lipid and alcohol metabolism. Five colon-related domains had genes for apoptosis, cell proliferation, ion transporter and mucin production. These clusters were very small and expression levels were low.

Brain and breast related genes didn't cluster. This shows that at least some clusters consist of functionally related genes. However, there are great exceptions. Spellman and Rubin have shown that there are clusters of co-expressed genes that are not functionally related. It seems like that clusters appear in very different forms.

Regulation

Cohen et al. found that of a pair of co-expressed genes only one promoter has an Upstream Activating Sequence (UAS) associated with that expression pattern. They suggested that UASs can activate genes that are not in immediate adjacency to them. This explanation could explain the co-expression of small clusters, but many clusters contain to many genes to be regulated by a single UAS. Chromatin changes are a plausible explanation for the co-regulation seen in clusters. Chromatin consists of the DNA strand and histones that are attached to the DNA. Regions were chromatin is very tightly packed are called heterochromatin. Heterochromatin consists very often of remains of viral genomes, transposons and other junk DNA. Because of tight packing the DNA is almost unreachable for the transcript machinery, covering deleterious DNA with proteins is the way in which the cell can protect itself. Chromatin which consists of functional genes is often an open structure were the DNA is accessible.

However, most of the genes are not needed to be expressed all the time. DNA with genes that aren't needed can be covered with histones. When a gene must be expressed special proteins can alter the chemical that are attached to the histones (histone modifications) that cause the histones to open the structure. When the chromatin of one gene is opened, the chromatin of the adjacent genes is also until this modification meets a boundary element. In that way genes is close proximity are expressed on the same time. So, genes are clustered in "expression hubs". In comparison with this model Gilbert et al. (2004) showed that RIDGEs are mostly present in open chromatin structures. However Johnidis et al. (2005) have shown that genes in the same cluster can be very differently expressed. How eukaryotic gene regulation, and associated chromatin changes, precisely works is still very unclear and there is no consensus about it. In order to get a clear picture about the mechanism of gene clusters first the workings chromatin and gene regulation needs to be illuminated. Furthermore most papers that identified clusters of co-regulated genes focused on transcription levels whereas few focused on clusters regulated by the same transcription-factors. Johnides et al. discovered strange phenomena when they did.

Origins

The fist models which tried to explain the clustering of genes were, of course, focused on operons because they were discovered before eukaryote gene clusters were. In 1999 Lawrence proposed a model for the origin operons. This selfish operon model suggests that individual genes were grouped together by vertical en horizontal transfer and were preserved as a single unit because that was beneficial for the genes, not per see for the organism. This model predicts that the gene clusters must have conserved between species. This is not the case for many operons and gene clusters seen in eukaryotes.

According to Eichler and Sankoff the two mean processes in eukaryotic chromosome evolution are 1) rearrangements of chromosomal segments and 2) localized duplication of genes. Clustering could be explained by reasoning that all genes in a cluster are originated from tandem duplicates of a common ancestor. If all co-expressed genes in a cluster were evolved from a common ancestral gene it would have been expected that they're co-expressed because they all have comparable promoters. However, gene clustering is a very common tread in genomes and it isn't clear how this duplication model could explain all of the clustering. Furthermore many genes that are present in clusters are not homologous.

How did evolutionary non-related genes come in close proximity in the first place? Either there is a force that brings functionally related genes near to each other, or the genes came near by change. Singer et al. proposed that genes came in close proximity by random recombination of genome segments. When functionally related genes came in close proximity to each other, this proximity was conserved. They determined all possible recombination sites between genes of human and mouse.

After that, they compared the clustering of the mouse and human genome and looked if recombination had occurred at the potentially recombination sites. It turned out that recombination between genes of the same cluster was very rare. So, as soon as a functional cluster is formed recombination is suppressed by the cell. On sex chromosomes, the amount of clusters is very low in both human and mouse. The authors reasoned this was due to the low rate of chromosomal rearrangements of sex chromosomes. Open chromatin regions are active regions. It is more likely that genes will be transferred to these regions. Genes from organelle and virus genome are inserted more often in these regions. In this way non-homologous genes can be pressed together in a small domain.

It is possible that some regions in the genome are better suited for important genes. It is important for the cell that genes that are responsible for basal functions are protected from recombination. It has been observed in yeast and worms that essential genes tent to cluster in regions with a small replication rate. It is possible that genes came in close proximity by change. Other models have been proposed but none of them can explain all observed phenomena. It's clear that as soon as clusters are formed they are conserved by natural selection. However, a precise model of how genes came in close proximity is still lacking.

The bulk of the present clusters must have formed relatively recent because only seven clusters of functionally related genes are conserved between phyla. Some of these differences can be explained by the fact that gene expression is very differently regulated by different phyla. For example, in vertebrates and plants DNA methylation is used, whereas it is absent in yeast and flies.

AlloMap Molecular Expression Testing

AlloMap molecular expression testing, developed and commercialized by XDx, is a gene expression profiling test to identify heart transplant recipients with a low probability of one type of transplant rejection. The test is performed on a blood sample, providing a non-invasive test to help manage the care of patients post transplant. Prior to the availability of this test, the primary method for managing heart transplant rejection was the invasive technique of endomyocardial biopsy. Test results are reported as a single score indicating the probability of moderate/severe acute cellular rejection (ACR). The performance characteristics of the test make it best suited to help indicate that acute cellar rejection is not present.

The score is based on the amount of RNA from each gene in a 20-gene panel comprising 11 rejection-related genes and 9 genes used for normalization and quality control. Many of the rejection-related genes are associated with biological pathways involved in the immune response and rejection processes. The test score is used, along with other standard clinical assessments, to evaluate the patient's probability of acute cellular rejection and the need for additional evaluations.

This test is not designed to be informative about other forms of heart rejections such as antibody-mediated rejection (AMR) or cardiac allograft vasculopathy (CAV). AlloMap has been commercially available since 2005 as a CLIA approved Laboratory Developed Test (LDT) and was cleared by the U.S. Food and Drug Administration (FDA) in 2008 as a Class II Medical Device. It is available only from the XDx Reference Laboratory in Brisbane, CA. The use of the test is described in the recommendations for the non-invasive monitoring of acute heart transplant rejection in the first evidence-based clinical practice guidelines for the care of heart transplant recipients issued by the International Society of Heart and Lung Transplantation.

Development

The test was developed using genomics and bioinformatics technologies. DNA microarrays were used to discover 252 candidate genes for which the amount of RNA in blood samples was related to rejection. Quantitative real-time polymerase chain reaction technology (qRT-PCR) confirmed 68 of the candidate genes from which the 20-gene gene expression panel was selected. The diagnostic performance was verified using independent patient samples from a multicenter clinical study. Initial clinical experience at three medical centers was published in 2006, confirming the efficacy and performance of the AlloMap test.

Clinical Studies

CARGO Study

The development and clinical validation of the test used patient samples and clinical data obtained during the Cardiac Allograft Rejection Gene Expression Observational (CARGO) Study. From 2001 to 2005, 737 patients from nine U.S. transplant centers enrolled in the Study and contributed 5,834 blood samples and associated clinical data. Initial clinical experience at three medical centers was published in 2006, confirming the efficacy and performance of the test.

IMAGE Study

A comparative effectiveness study, the Invasive Monitoring Attenuation through Gene Expression (IMAGE) Study, compared clinical outcomes of patients managed with AlloMap to outcomes of patients managed with endomyocardial biopsy. The study, which ran from 2005–09, included 602 patients from thirteen U.S. centers who were at least six months post-transplant. The results showed that AlloMap was not inferior to endomyocardial biopsy with respect to clinical outcomes when used to monitor stable, asymptomatic heart transplant patients.

Indications for Use

The test is currently indicated for use in heart transplant recipients 15 years of age or older, and at least 2 months (e”55 days) post-transplant.

Method of Use

The test is based on standard quantitative real-time polymerase chain reaction technology (qRT-PCR) using RNA isolated from peripheral blood mononuclear cells (PBMC). A blood sample is collected, PBMC are isolated, lysed and the released RNA stabilized and frozen (PBMC lysate). RNA is then purified from the PBMC lysate, converted into complementary DNA (cDNA), and mixed with gene-specific primers and probes. The expression of each gene is measured by amplification and fluorescence detection using a qRT-PCR instrument. A mathematical classifier combines the measured expression values for each gene into a single value reported as a score between 0 and 40. Each score is associated with a negative predictive value (NPV) and a positive predictive value (PPV). The test is characterized by high negative predictive values and is therefore a test used to help identify patients at low probability of rejection. The test has a relatively low positive predictive value, meaning that even when the score is relatively high, the risk of rejection may still be low.

2

Organisation of Genome

Genome organization refers to the sequential, not the structural organization of the genome. Besides the coding exons, the non-coding DNA in Eukaryotes may fall in the following classes:

- Introns. They are DNA sequences inserted between the exons and found in the ORF. They are spliced after the first level of transcription. Most introns are junk inserted within genes.
- Pseudogenes. 'Dead', non-functional copies of genes present elsewhere in the genome, but no longer of any use.
- Retropseudogenes. Like pseudogenes, but have been processed, i.e. lack introns. Produced by the action of reverse transcriptase (RT) on mRNA, and subsequent incorporation of the cDNA into the genome.
- Transposons. Jumping genes, which splice themselves in and out of the genome (in DNA form) randomly, by the action of transposase.
- Retrotransposons. Transcribed into an mRNA, which encodes an RT enzyme, which then copies the mRNA back to DNA and incorporates it into the genome.

Infact in humans only 1.5% of the entire genome length corresponds to coding DNA. This 1.5% codes for about 27,000 genes which in turn code for proteins that are responsible for all the cellular processes.

Genome Analysis?

Genome analysis entails the prediction of genes in uncharacterized genomic sequences. The 21st century has seen the announcement of the draft version of the human genome sequence. Model organisms have been sequenced in both the plant and animal kingdoms. However, the pace of genome annotation is not matching the pace of genome sequencing. Experimental genome annotation is slow and time consuming. The demand is to be able to develop computational tools for gene prediction.

Computational Gene prediction is relatively simple for the prokaryotes where all the genes are converted into the corresponding mRNA and then into proteins. The process is more complex for eukaryotic cells where the coding DNA sequence is interrupted by random sequences called introns.

Some of the questions which biologists want to answer today are:

- Given a DNA sequence, what part of it codes for a protein and what part of it is junk DNA.
- Classify the junk DNA as intron, untranslated region, transposons, dead genes, regulatory elements etc.
- Divide a newly sequenced genome into the genes (coding) and the non-coding regions.

The importance of genome analysis can be understood by comparing the human and chimpanzee genomes. The chimp and human genomes vary by an average of just 2% i.e. just about 160 enzymes. A complete genome analysis of the two genomes would give a strong insight into the various mechanisms responsible for the differences.Arabidopsis and Humans have the same number of genes, though the Arabidopsis genome is around 250 times smaller than humans. How is that ?

The human genome has a lot of junk DNA, specifically transposons and mobile genetic elements. This increases the size of the human genome, though the number of genes is only 27,000.

However, the number of protein products in humans is significantly higher. Many of the sequenced human genes have alternative splice products. In addition, several other processes (e.g. signal transduction) proceed via further protein modifications, such as Glycosylation. Therefore, the number of human protein products could far exceed the number of genes. Why do plants have such bulky genomes when they are not as complex as some of the higher eukaryotes ? This is mostly due to two factors: the ability of plants to duplicate their genomes in order to reproduce (a process known as polyploidization) and the susceptibility of plants to mobile genetic elements.

- Polyploidization allows plants to more easily form hybrids when pollen and ova from different species come together. The result of such hybridization events are plants with genomes that are the sum of the two parent genome sizes (as opposed to half of one parent's genome and half of the other parent's genome as in normal sexual reproduction.

- Also, in case of plants, it is fairly common to observe insertion of transposable elements in intergenic regions. This also explains the difference in the sizes of plant genomes among themselves as well.

What is Open Reading Frame (ORF) ?

The region of the nucleotide sequences from the start codon (ATG) to the stop codon is called the Open Reading frame. Gene finding in organism specially prokaryotes starts form searching for an open reading frames (ORF). An ORF is a sequence of DNA that starts with start codon "ATG" (not always) and ends with any of the three termination codons (TAA, TAG, TGA).

Depending on the starting point, there are six possible ways (three on forward strand and three on complementary strand) of translating any nucleotide sequence into amino acid sequence according to the genetic code.These are called reading frames. While eukaryotic gene finding is altogether a different task as the eukaryotic genes are not continuous and interrupted by intervening noncoding sequences called 'introns'. Moreover organization of genetic information in eukaryotes and prokaryotes is different.

What is Coding Sequence(CDS)? How is it Different from the ORF?

The Coding Sequence (CDS) is the actual region of DNA that is translated to form proteins. While the ORF may contain introns as well, the CDS refers to those nucleotides(concatenated exons) that can be divided into codons which are actually translated into amino acids by the ribosomal translation machinery. In Prokaryotes the ORF and the CDS are the same.

DNA Replication

Before a cell divides, its DNA is replicated (duplicated.) Because the two strands of a DNA molecule have complementary base pairs, the nucleotide sequence of each strand automatically supplies the information needed to produce its partner. If the two strands of a DNA molecule are separated, each can be used as a pattern or template to produce a complementary strand. Each template and its new complement together then form a new DNA double helix, identical to the original.

Before replication can occur, the length of the DNA double helix about to be copied must be unwound. In addition, the two strands must be separated, much like the two sides of a zipper, by breaking the weak hydrogen bonds that link the paired bases. Once the DNA strands have been unwound, they must be held apart to expose the bases so that new nucleotide partners can hydrogen-bond to them.

The enzyme DNA polymerase then moves along the exposed DNA strand, joining newly arrived nucleotides into a new DNA strand that is complementary to the template. Each cell contains a family of more than thirty enzymes to insure the accurate replication of DNA.

Primers

Though DNA polymerase can elongate a polynucleotide strand by adding new nucleotides, it cannot start a strand from scratch because it can only bond new nucleotides to a free sugar (3') end of a nucleotide chain. DNA polymerase requires the assistance of a primer, a previously existing short strand of DNA (or RNA) that is complementary to the first part of the DNA segment being copied. This small strand of nucleotides anneals (binds) by complementary base pairing to the beginning of the area being copied. With the primer in place, DNA polymerase is then able to continue adding the rest of the pairs of the segment until a new double strand of DNA is completed. Primers are formed from free nucleotides in the cell by enzymes called DNA primases.

Replication Occurs Differently on Antiparallel Strands of DNA

That nucleotides can be added only to the sugar or 3' end of the growing complementary chain presents no problem for the side of the DNA chain opening at its phosphate or 5' end. The primer that binds to the first few exposed bases will end with a sugar (3') where the phosphate of a new nucleotide can be attached. From there on, DNA polymerase can continuously synthesize the growing complementary strand. This strand of DNA is called the leading strand. A different challenge faces DNA polymerase when the complementary side of the DNA molecule begins unzipping from its sugar (3') toward its phosphate (5') end. A primer of complementary molecules attaching to the opening end of this chain would have a phosphate not a sugar at its exposed end so that new nucleotides could not be joined.

To get around this problem, this strand is synthesized in small pieces backward from the overall direction of replication. This strand is called the lagging strand. The short segments of newly assembled DNA from which the lagging strand is built are called Okazaki fragments. As replication proceeds and nucleotides are added to the 3' end of the Okazaki fragments, they come to meet each other. The primer fragments are then booted out by enzymes and replaced by appropriate DNA nucleotides. The whole thing is then stitched together by another enzyme called DNA ligase.

DNA Damage, Repair, and Recombination

DNA repair refers to a collection of processes by which a cell identifies and corrects damage to the DNA molecules that encode its genome. In human cells, both normal metabolic activities and environmental factors such as UV light and radiation can cause DNA damage, resulting in as many as 1 million individual molecular lesions per cell per day. Many of these lesions cause structural damage to the DNA molecule and can alter or eliminate the cell's ability to transcribe the gene that the affected DNA encodes. Other lesions induce potentially harmful mutations in the cell's genome, which affect the survival of its daughter cells after it undergoes mitosis. As a consequence, the DNA repair process is constantly active as it responds to damage in the DNA structure. When normal repair processes fail, and when cellular apoptosis does not occur, irreparable DNA damage may occur, including double-strand breaks and DNA crosslinkages.

The rate of DNA repair is dependent on many factors, including the cell type, the age of the cell, and the extracellular environment. A cell that has accumulated a large amount of DNA damage, or one that no longer effectively repairs damage incurred to its DNA, can enter one of three possible states:

1. an irreversible state of dormancy, known as senescence
2. cell suicide, also known as apoptosis or programmed cell death
3. unregulated cell division, which can lead to the formation of a tumor that is cancerous.

The DNA repair ability of a cell is vital to the integrity of its genome and thus to its normal functioning and that of the organism. Many genes that were initially shown to influence life span have turned out to be involved in DNA damage repair and protection. Failure to correct molecular lesions in cells that form gametes can introduce mutations into the genomes of the offspring and thus influence the rate of evolution.

DNA Damage

DNA damage, due to environmental factors and normal metabolic processes inside the cell, occurs at a rate of 1,000 to 1,000,000 molecular lesions per cell per day. While this constitutes only 0.000165% of the human genome's approximately 6 billion bases (3 billion base pairs), unrepaired lesions in critical genes (such as tumor suppressor genes) can impede a cell's ability to carry out its function and appreciably increase the likelihood of tumour formation.

The vast majority of DNA damage affects the primary structure of the double helix; that is, the bases themselves are chemically modified. These modifications can in turn disrupt the molecules' regular helical structure by introducing non-native chemical bonds or bulky adducts that do not fit in the standard double helix. Unlike proteins and RNA, DNA usually lacks tertiary structure and therefore damage or disturbance does not occur at that level. DNA is, however, supercoiled and wound around "packaging" proteins called histones (in eukaryotes), and both superstructures are vulnerable to the effects of DNA damage.

Sources of Damage

DNA damage can be subdivided into two main types:

1. endogenous damage such as attack by reactive oxygen species produced from normal metabolic byproducts (spontaneous mutation), especially the process of oxidative deamination
   - also includes replication errors.
2. exogenous damage caused by external agents such as
   - ultraviolet [UV 200-300nm] radiation from the sun
   - other radiation frequencies, including x-rays and gamma rays
   - hydrolysis or thermal disruption
   - certain plant toxins
   - human-made mutagenic chemicals, especially aromatic compounds that act as DNA intercalating agents
   - cancer chemotherapy and radiotherapy
   - viruses.

The replication of damaged DNA before cell division can lead to the incorporation of wrong bases opposite damaged ones. Daughter cells that inherit these wrong bases carry mutations from which the original DNA sequence is unrecoverable (except in the rare case of a back mutation, for example, through gene conversion).

Types of Damage

There are five main types of damage to DNA due to endogenous cellular processes:

1. *oxidation* of bases [e.g. 8-oxo-7,8-dihydroguanine (8-oxoG)] and generation of DNA strand interruptions from reactive oxygen species,
2. *alkylation* of bases (usually methylation), such as formation of 7-methylguanine, 1-methyladenine, 6-O-Methylguanine
3. *hydrolysis* of bases, such as deamination, depurination, and depyrimidination.
4. "bulky adduct formation" (i.e., benzo[a]pyrene diol epoxide-dG adduct)
5. *mismatch* of bases, due to errors in DNA replication, in which the wrong DNA base is stitched into place in a newly forming DNA strand, or a DNA base is skipped over or mistakenly inserted.

Damage caused by exogenous agents comes in many forms. Some examples are:

1. *UV-B light* causes crosslinking between adjacent cytosine and thymine bases creating *pyrimidine dimers*. This is called direct DNA damage.

2. *UV-A light* creates mostly free radicals. The damage caused by free radicals is called indirect DNA damage.
3. *Ionizing radiation* such as that created by radioactive decay or in *cosmic rays* causes breaks in DNA strands. Low-level ionizing radiation may induce irreparable DNA damage (leading to replicational and transcriptional errors needed for neoplasia or may trigger viral interactions) leading to pre-mature aging and cancer.
4. *Thermal disruption* at elevated temperature increases the rate of depurination (loss of purine bases from the DNA backbone) and single-strand breaks. For example, hydrolytic depurination is seen in the thermophilic bacteria, which grow in hot springs at 40-80 °C. The rate of depurination (300 purine residues per genome per generation) is too high in these species to be repaired by normal repair machinery, hence a possibility of an adaptive response cannot be ruled out.
5. *Industrial chemicals* such as vinyl chloride and hydrogen peroxide, and environmental chemicals such as polycyclic hydrocarbons found in smoke, soot and tar create a huge diversity of DNA adducts- ethenobases, oxidized bases, alkylated phosphotriesters and Crosslinking of DNA just to name a few.

UV damage, alkylation/methylation, X-ray damage and oxidative damage are examples of induced damage. Spontaneous damage can include the loss of a base, deamination, sugar ring puckering and tautomeric shift.

Nuclear versus Mitochondrial DNA Damage

In human cells, and eukaryotic cells in general, DNA is found in two cellular locations - inside the nucleus and inside the mitochondria. Nuclear DNA (nDNA) exists as chromatin during non-replicative stages of the cell cycle and is condensed into aggregate structures known as chromosomes during cell division. In either state the DNA is highly compacted and wound up around bead-like proteins called histones.

Whenever a cell needs to express the genetic information encoded in its nDNA the required chromosomal region is unravelled, genes located therein are expressed, and then the region is condensed back to its resting conformation. Mitochondrial DNA (mtDNA) is located inside mitochondria organelles, exists in multiple copies, and is also tightly associated with a number of proteins to form a complex known as the nucleoid. Inside mitochondria, reactive oxygen species (ROS), or free radicals, byproducts of the constant production of adenosine triphosphate (ATP) via oxidative phosphorylation, create a highly oxidative environment that is known to damage mtDNA. A critical enzyme in counteracting the toxicity of these species is superoxide dismutase, which is present in both the mitochondria and cytoplasm of eukaryotic cells.

Senescence and Apoptosis

Senescence, an irreversible state in which the cell no longer divides, is a protective response to the shortening of the chromosome ends. The telomeres are long regions of repetitive noncoding DNA that cap chromosomes and undergo partial degradation each time a cell undergoes division. In contrast, quiescence is a reversible state of cellular dormancy that is unrelated to genome damage. Senescence in cells may serve as a functional alternative to apoptosis in cases where the physical presence of a cell for spatial reasons is required by the organism, which serves as a "last resort" mechanism to prevent a cell with damaged DNA from replicating inappropriately in the absence of pro-growth cellular signaling. Unregulated cell division can lead to the formation of a tumour, which is potentially lethal to an organism. Therefore, the induction of senescence and apoptosis is considered to be part of a strategy of protection against cancer.

DNA Damage and Mutation

It is important to distinguish between DNA damage and mutation, the two major types of error in DNA. DNA damages and mutation are fundamentally different. Damages are physical abnormalities in the DNA, such as single- and double-strand breaks, 8-hydroxydeoxyguanosine residues, and polycyclic aromatic hydrocarbon adducts. DNA damages can be recognized by enzymes, and, thus, they can be correctly repaired if redundant information, such as the undamaged sequence in the complementary DNA strand or in a homologous chromosome, is available for copying. If a cell retains DNA damage, transcription of a gene can be prevented, and, thus, translation into a protein will also be blocked. Replication may also be blocked and/or the cell may die.

In contrast to DNA damage, a mutation is a change in the base sequence of the DNA. A mutation cannot be recognized by enzymes once the base change is present in both DNA strands, and, thus, a mutation cannot be repaired. At the cellular level, mutations can cause alterations in protein function and regulation. Mutations are replicated when the cell replicates. In a population of cells, mutant cells will increase or decrease in frequency according to the effects of the mutation on the ability of the cell to survive and reproduce. Although distinctly different from each other, DNA damages and mutations are related because DNA damages often cause errors of DNA synthesis during replication or repair; these errors are a major source of mutation.

Given these properties of DNA damage and mutation, it can be seen that DNA damages are a special problem in non-dividing or slowly dividing cells, where unrepaired damages will tend to accumulate over time. On the other hand, in rapidly dividing cells, unrepaired DNA damages that do not kill the cell by blocking replication will tend to cause replication errors and thus mutation. The great majority of mutations that are not neutral in their effect are deleterious to a cell's survival. Thus, in a population of cells comprising a tissue with replicating cells, mutant cells will tend to be lost. However, infrequent mutations that provide a survival advantage will tend to clonally expand at the expense of neighboring cells in the tissue. This advantage to the cell is disadvantageous to the whole organism, because such mutant cells can give rise to cancer. Thus, DNA damages in frequently dividing cells, because they give rise to mutations, are a prominent cause of cancer. In contrast, DNA damages in infrequently dividing cells are likely a prominent cause of aging.

DNA Repair Mechanisms

Cells cannot function if DNA damage corrupts the integrity and accessibility of essential information in the genome (but cells remain superficially functional when so-called "non-essential" genes are missing or damaged). Depending on the type of damage inflicted on the DNA's double helical structure, a variety of repair strategies have evolved to restore lost information. If possible, cells use the unmodified complementary strand of the DNA or the sister chromatid as a template to recover the original information. Without access to a template, cells use an error-prone recovery mechanism known as translesion synthesis as a last resort. Damage to DNA alters the spatial configuration of the helix, and such alterations can be detected by the cell. Once damage is localized, specific DNA repair molecules bind at or near the site of damage, inducing other molecules to bind and form a complex that enables the actual repair to take place.

Direct Reversal

Cells are known to eliminate three types of damage to their DNA by chemically reversing it. These mechanisms do not require a template, since the types of damage they counteract can occur in only one of the four bases. Such direct reversal mechanisms are specific to the type of damage incurred and do not involve breakage of the phosphodiester backbone.

The formation of pyrimidine dimers upon irradiation with UV light results in an abnormal covalent bond between adjacent pyrimidine bases. The photoreactivation process directly reverses this damage by the action of the enzyme photolyase, whose activation is obligately dependent on energy absorbed from blue/UV light (300–500 nm wavelength) to promote catalysis. Another type of damage, methylation of guanine bases, is directly reversed by the protein methyl guanine methyl transferase (MGMT), the bacterial equivalent of which is called ogt.

This is an expensive process because each MGMT molecule can be used only once; that is, the reaction is stoichiometric rather than catalytic. A generalized response to methylating agents in bacteria is known as the adaptive response and confers a level of resistance to alkylating agents upon sustained exposure by upregulation of alkylation repair enzymes. The third type of DNA damage reversed by cells is certain methylation of the bases cytosine and adenine.

Single-strand Damage

When only one of the two strands of a double helix has a defect, the other strand can be used as a template to guide the correction of the damaged strand. In order to repair damage to one of the two paired molecules of DNA, there exist a number of excision repair mechanisms that remove the damaged nucleotide and replace it with an undamaged nucleotide complementary to that found in the undamaged DNA strand.

1. Base excision repair (BER), which repairs damage to a single base caused by oxidation, alkylation, hydrolysis, or deamination. The damaged base is removed by a DNA glycosylase. The "missing tooth" is then recognised by an enzyme called AP endonuclease, which cuts the Phosphodiester bond. The missing part is then resynthesized by a DNA polymerase, and a DNA ligase performs the final nick-sealing step.
2. Nucleotide excision repair (NER), which recognizes bulky, helix-distorting lesions such as pyrimidine dimers and 6,4 photoproducts. A specialized form of NER known as transcription-coupled repair deploys NER enzymes to genes that are being actively transcribed.
3. Mismatch repair (MMR), which corrects errors of DNA replication and recombination that result in mispaired (but undamaged) nucleotides.

Double-strand Breaks

Double-strand breaks, in which both strands in the double helix are severed, are particularly hazardous to the cell because they can lead to genome rearrangements. Three mechanisms exist to repair double-strand breaks (DSBs): non-homologous end joining (NHEJ), microhomology-mediated end joining (MMEJ), and homologous recombination. PVN Acharya noted that double-strand breaks and a "cross-linkage joining both strands at the same point is irreparable because neither strand can then serve as a template for repair. The cell will die in the next mitosis or in some rare instances, mutate."

In NHEJ, DNA Ligase IV, a specialized DNA ligase that forms a complex with the cofactor XRCC4, directly joins the two ends. To guide accurate repair, NHEJ relies on short homologous sequences called microhomologies present on the single-stranded tails of the DNA ends to be joined.

If these overhangs are compatible, repair is usually accurate. NHEJ can also introduce mutations during repair. Loss of damaged nucleotides at the break site can lead to deletions, and joining of nonmatching termini forms translocations. NHEJ is especially important before the cell has replicated its DNA, since there is no template available for repair by homologous recombination. There are "backup" NHEJ pathways in higher eukaryotes. Besides its role as a genome caretaker, NHEJ is required for joining hairpin-capped double-strand breaks induced during V(D)J recombination, the process that generates diversity in B-cell and T-cell receptors in the vertebrate immune system.

Homologous recombination requires the presence of an identical or nearly identical sequence to be used as a template for repair of the break. The enzymatic machinery responsible for this repair process is nearly identical to the machinery responsible for chromosomal crossover during meiosis. This pathway allows a damaged chromosome to be repaired using a sister chromatid (available in G2 after DNA replication) or a homologous chromosome as a template. DSBs caused by the replication machinery attempting to synthesize across a single-strand break or unrepaired lesion cause collapse of the replication fork and are typically repaired by recombination. Topoisomerases introduce both single- and double-strand breaks in the course of changing the DNA's state of supercoiling, which is especially common in regions near an open replication fork. Such breaks are not considered DNA damage because they are a natural intermediate in the topoisomerase biochemical mechanism and are immediately repaired by the enzymes that created them. A team of French researchers bombarded *Deinococcus radiodurans* to study the mechanism of double-strand break DNA repair in that organism. At least two copies of the genome, with random DNA breaks, can form DNA fragments through annealing. Partially overlapping fragments are then used for synthesis of homologous regions through a moving D-loop that can continue extension until they find complementary partner strands. In the final step there is crossover by means of RecA-dependent homologous recombination.

Translesion Synthesis

Translesion synthesis is a DNA damage tolerance process that allows the DNA replication machinery to replicate past DNA lesions such as thymine dimers or AP sites. It involves switching out regular DNA polymerases for specialized translesion polymerases (i.e. DNA polymerase IV or V, from the Y Polymerase family), often with larger active sites that can facilitate the insertion of bases opposite damaged nucleotides. The polymerase switching is thought to be mediated by, among other factors, the post-translational modification of the replication processivity factor PCNA. Translesion synthesis polymerases often have low fidelity (high propensity to insert wrong bases) on undamaged templates relative to regular polymerases. However, many are extremely efficient at inserting correct bases opposite specific types of damage.

For example, Pol ç mediates error-free bypass of lesions induced by UV irradiation, whereas Pol æ introduces mutations at these sites. From a cellular perspective, risking the introduction of point mutations during translesion synthesis may be preferable to resorting to more drastic mechanisms of DNA repair, which may cause gross chromosomal aberrations or cell death. In short, the process involves specialized polymerases either bypassing or repairing lesions at locations of stalled DNA replication.

A bypass platform is provided to these polymerases by Proliferating cell nuclear antigen (PCNA). Under normal circumstances, PCNA bound to polymerases replicates the DNA. At a site of lesion, PCNA is ubiquitinated, or modified, by the RAD6/RAD18 proteins to provide a platform for the specialized polymerases to bypass the lesion and resume DNA replication.

Global Response to DNA Damage

Cells exposed to ionizing radiation, ultraviolet light or chemicals are prone to acquire multiple sites of bulky DNA lesions and double-strand breaks. Moreover, DNA damaging agents can damage other biomolecules such as proteins, carbohydrates, lipids, and RNA. The accumulation of damage, to be specific, double-strand breaks or adducts stalling the replication forks, are among known stimulation signals for a global response to DNA damage. The global response to damage is an act directed toward the cells' own preservation and triggers multiple pathways of macromolecular repair, lesion bypass, tolerance, or apoptosis. The common features of global response are induction of multiple genes, cell cycle arrest, and inhibition of cell division.

DNA Damage Checkpoints

After DNA damage, cell cycle checkpoints are activated. Checkpoint activation pauses the cell cycle and gives the cell time to repair the damage before continuing to divide. DNA damage checkpoints occur at the G1/S and G2/M boundaries. An intra-S checkpoint also exists. Checkpoint activation is controlled by two master kinases, ATM and ATR. ATM responds to DNA double-strand breaks and disruptions in chromatin structure, whereas ATR primarily responds to stalled replication forks. These kinases phosphorylate downstream targets in a signal transduction cascade, eventually leading to cell cycle arrest. A class of checkpoint mediator proteins including BRCA1, MDC1, and 53BP1 has also been identified. These proteins seem to be required for transmitting the checkpoint activation signal to downstream proteins.

P53 is an important downstream target of ATM and ATR, as it is required for inducing apoptosis following DNA damage. At the G1/S checkpoint, p53 functions by deactivating the CDK2/cyclin E complex. Similarly, p21 mediates the G2/M checkpoint by deactivating the CDK1/cyclin B complex.

The Prokaryotic SOS Response

The SOS response is the term used to describe changes in gene expression in *Escherichia coli* and other bacteria in response to extensive DNA damage. The prokaryotic SOS system is regulated by two key proteins: LexA and RecA. The LexA homodimer is a transcriptional repressor that binds to operator sequences commonly referred to as SOS boxes.

In *Escherichia coli* it is known that LexA regulates transcription of approximately 48 genes including the lexA and recA genes. The SOS response is known to be widespread in the Bacteria domain, but it is mostly absent in some bacterial phyla, like the Spirochetes.

The most common cellular signals activating the SOS response are regions of single-stranded DNA (ssDNA), arising from stalled replication forks or double-strand breaks, which are processed by DNA helicase to separate the two DNA strands. In the initiation step, RecA protein binds to ssDNA in an ATP hydrolysis driven reaction creating RecA–ssDNA filaments. RecA–ssDNA filaments activate LexA autoprotease activity, which ultimately leads to cleavage of LexA dimer and subsequent LexA degradation. The loss of LexA repressor induces transcription of the SOS genes and allows for further signal induction, inhibition of cell division and an increase in levels of proteins responsible for damage processing.

In *Escherichia coli*, SOS boxes are 20-nucleotide long sequences near promoters with palindromic structure and a high degree of sequence conservation. In other classes and phyla, the sequence of SOS boxes varies considerably, with different length and composition, but it is always highly conserved and one of the strongest short signals in the genome. The high information content of SOS boxes permits differential binding of LexA to different promoters and allows for timing of the SOS response. The lesion repair genes are induced at the beginning of SOS response. The error-prone translesion polymerases, for example, UmuCD'2 (also called DNA polymerase V), are induced later on as a last resort. Once the DNA damage is repaired or bypassed using polymerases or through recombination, the amount of single-stranded DNA in cells is decreased, lowering the amounts of RecA filaments decreases cleavage activity of LexA homodimer, which then binds to the SOS boxes near promoters and restores normal gene expression.

Eukaryotic Transcriptional Responses to DNA Damage

Eukaryotic cells exposed to DNA damaging agents also activate important defensive pathways by inducing multiple proteins involved in DNA repair, cell cycle checkpoint control, protein trafficking and degradation. Such genome wide transcriptional response is very complex and tightly regulated, thus allowing coordinated global response to damage. Exposure of yeast

Saccharomyces cerevisiae to DNA damaging agents results in overlapping but distinct transcriptional profiles. Similarities to environmental shock response indicates that a general global stress response pathway exist at the level of transcriptional activation. In contrast, different human cell types respond to damage differently indicating an absence of a common global response. The probable explanation for this difference between yeast and human cells may be in the heterogeneity of mammalian cells. In an animal different types of cells are distributed among different organs that have evolved different sensitivities to DNA damage.

In general global response to DNA damage involves expression of multiple genes responsible for postreplication repair, homologous recombination, nucleotide excision repair, DNA damage checkpoint, global transcriptional activation, genes controlling mRNA decay, and many others. A large amount of damage to a cell leaves it with an important decision: undergo apoptosis and die, or survive at the cost of living with a modified genome. An increase in tolerance to damage can lead to an increased rate of survival that will allow a greater accumulation of mutations. Yeast Rev1 and human polymerase ç are members of [Y family translesion DNA polymerases present during global response to DNA damage and are responsible for enhanced mutagenesis during a global response to DNA damage in eukaryotes.

DNA Repair and Aging

Pathological Effects of Poor DNA Repair

DNA repair rate is an important determinant of cell pathology. Experimental animals with genetic deficiencies in DNA repair often show decreased life span and increased cancer incidence. For example, mice deficient in the dominant NHEJ pathway and in telomere maintenance mechanisms get lymphoma and infections more often, and, as a consequence, have shorter lifespans than wild-type mice. In similar manner, mice deficient in a key repair and transcription protein that unwinds DNA helices have premature onset of aging-related diseases and consequent shortening of lifespan. However, not every DNA repair deficiency creates exactly the predicted effects; mice deficient in the NER pathway exhibited shortened life span without correspondingly higher rates of mutation.

If the rate of DNA damage exceeds the capacity of the cell to repair it, the accumulation of errors can overwhelm the cell and result in early senescence, apoptosis, or cancer. Inherited diseases associated with faulty DNA repair functioning result in premature aging, increased sensitivity to carcinogens, and correspondingly increased cancer risk. On the other hand, organisms with enhanced DNA repair systems, such as *Deinococcus radiodurans*, the most radiation-resistant known organism, exhibit remarkable resistance to the double-strand break-inducing effects of radioactivity, likely due to enhanced efficiency of DNA repair and especially NHEJ.

Longevity and Caloric Restriction

A number of individual genes have been identified as influencing variations in life span within a population of organisms. The effects of these genes is strongly dependent on the environment, in particular, on the organism's diet. Caloric restriction reproducibly results in extended lifespan in a variety of organisms, likely via nutrient sensing pathways and decreased metabolic rate.

The molecular mechanisms by which such restriction results in lengthened lifespan are as yet unclear; however, the behavior of many genes known to be involved in DNA repair is altered under conditions of caloric restriction.

For example, increasing the gene dosage of the gene SIR-2, which regulates DNA packaging in the nematode worm *Caenorhabditis elegans*, can significantly extend lifespan. The mammalian homolog of SIR-2 is known to induce downstream DNA repair factors involved in NHEJ, an activity that is especially promoted under conditions of caloric restriction. Caloric restriction has been closely linked to the rate of base excision repair in the nuclear DNA of rodents, although similar effects have not been observed in mitochondrial DNA.

It is interesting to note that the *C. elegans* gene AGE-1, an upstream effector of DNA repair pathways, confers dramatically extended life span under free-feeding conditions but leads to a decrease in reproductive fitness under conditions of caloric restriction. This observation supports the pleiotropy theory of the biological origins of aging, which suggests that genes conferring a large survival advantage early in life will be selected for even if they carry a corresponding disadvantage late in life.

Medicine and DNA Repair Modulation

Hereditary DNA Repair Disorders

Defects in the NER mechanism are responsible for several genetic disorders, including:

- Xeroderma pigmentosum: hypersensitivity to sunlight/UV, resulting in increased skin cancer incidence and premature aging
- Cockayne syndrome: hypersensitivity to UV and chemical agents
- Trichothiodystrophy: sensitive skin, brittle hair and nails.

Mental retardation often accompanies the latter two disorders, suggesting increased vulnerability of developmental neurons.

Other DNA repair disorders include:

- Werner's syndrome: premature aging and retarded growth

- Bloom's syndrome: sunlight hypersensitivity, high incidence of malignancies (especially leukemias).
- Ataxia telangiectasia: sensitivity to ionizing radiation and some chemical agents.

All of the above diseases are often called "segmental progerias" ("accelerated aging diseases") because their victims appear elderly and suffer from aging-related diseases at an abnormally young age, while not manifesting all the symptoms of old age. Other diseases associated with reduced DNA repair function include Fanconi's anemia, hereditary breast cancer and hereditary colon cancer.

DNA Repair and Cancer

Inherited mutations that affect DNA repair genes are strongly associated with high cancer risks in humans. Hereditary nonpolyposis colorectal cancer (HNPCC) is strongly associated with specific mutations in the DNA mismatch repair pathway. BRCA1 and BRCA2, two famous mutations conferring a hugely increased risk of breast cancer on carriers, are both associated with a large number of DNA repair pathways, especially NHEJ and homologous recombination.

Cancer therapy procedures such as chemotherapy and radiotherapy work by overwhelming the capacity of the cell to repair DNA damage, resulting in cell death. Cells that are most rapidly dividing - most typically cancer cells - are preferentially affected. The side-effect is that other non-cancerous but rapidly dividing cells such as stem cells in the bone marrow are also affected. Modern cancer treatments attempt to localize the DNA damage to cells and tissues only associated with cancer, either by physical means (concentrating the therapeutic agent in the region of the tumour) or by biochemical means (exploiting a feature unique to cancer cells in the body).

3

The Nature of the Gene

A gene is a molecular unit of heredity of a living organism. It is a name given to some stretches of DNA and RNA that code for a type of protein or for an RNA chain that has a function in the organism. Living beings depend on genes, as they specify all proteins and functional RNA chains. Genes hold the information to build and maintain an organism's cells and pass genetic traits to offspring, although some organelles (e.g. mitochondria) are self-replicating and are not coded for by the organism's DNA. All organisms have many genes corresponding to various different biological traits, some of which are immediately visible, such as eye colour or number of limbs, and some of which are not, such as blood type or increased risk for specific diseases, or the thousands of basic biochemical processes that comprise life. A modern working definition of a gene is "*a locatable region of genomic sequence, corresponding to a unit of inheritance, which is associated with regulatory regions, transcribed regions, and or other functional sequence regions* ". Colloquial usage of the term *gene* (e.g. "good genes", "hair colour gene") may actually refer to an allele: a *gene* is the basic instruction — a sequence of nucleic acids (DNA or, in the case of certain viruses RNA), while an *allele* is one variant of that gene. Referring to having a gene for a trait is no longer the scientifically accepted usage. In most cases, all people would have a gene for the trait in question, but certain people will have a specific allele of that gene, which results in the trait variant. Further, genes code for proteins, which might result in identifiable traits, but it is the gene, not the trait, which is inherited.

Genes are working subunits of DNA. DNA is a vast chemical information database that carries the complete set of instructions for making all the proteins a cell will ever need. Each gene contains a particular set of instructions, usually coding for a particular protein.

DNA exists as two long, paired strands spiraled into the famous double helix. Each strand is made up of millions of chemical building blocks called bases. While there are only four different chemical bases in DNA (adenine, thymine, cytosine, and guanine), the order in which the bases occur determines the information available, much as specific letters of the alphabet combine to form words and sentences.

DNA resides in the core, or nucleus, of each of the body's trillions of cells. Every human cell (with the exception of mature red blood cells, which have no nucleus) contains the same DNA. Each cell has 46 molecules of double-stranded DNA. Each molecule is made up of 50 to 250 million bases housed in a chromosome. The DNA in each chromosome constitutes many genes (as well as vast stretches of noncoding DNA, the function of which is unknown).

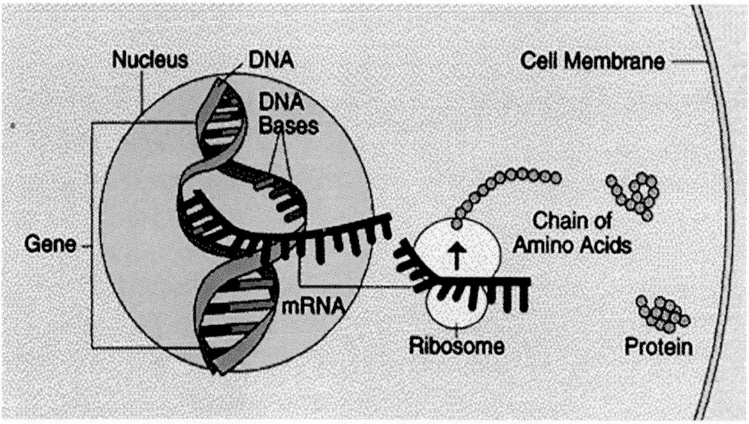

Figure : *For a cell to make protein, the information from a gene is copied, base by base, from DNA into new strands of messenger RNA (mRNA). Then mRNA travels out of the nucleus into the cytoplasm, to cell organelles called ribosomes. There, mRNA directs the assembly of amino acids that fold into completed protein molecule.*

A gene is any given segment along the DNA that encodes instructions that allow a cell to produce a specific product - typically, a protein such as an enzyme - that initiates one specific action. There are between 50,000 and 100,000 genes, and every gene is made up of thousands, even hundreds of thousands, of chemical bases.

Human cells contain two sets of chromosomes, one set inherited from the mother and one from the father. (Mature sperm and egg cells carry a single set of chromosomes.) Each set has 23 single chromosomes - 22 autosomes and an X or Y sex chromosome. (Females inherit an X from each parent, while males get an X from the mother and a Y from the father.)

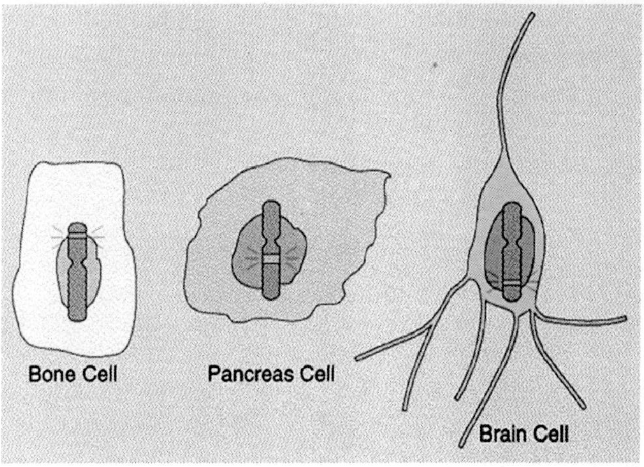

Figure : Different genes are activated in different cells, creating the specific proteins that give a particular cell type its character.

Physical Definitions

RNA Genes and Genomes

When proteins are manufactured, the gene is first copied into RNA as an intermediate product. In other cases, the RNA molecules are the actual functional products. For example, RNAs known as ribozymes are capable of enzymatic function, and microRNA has a regulatory role. The DNA sequences from which such RNAs are transcribed are known as RNA genes.

Some viruses store their entire genomes in the form of RNA, and contain no DNA at all. Because they use RNA to store genes, their cellular hosts may synthesize their proteins as soon as they are infected and without the delay in waiting for transcription. On the other hand, RNA retroviruses, such as HIV, require the reverse transcription of their genome from RNA into DNA before their proteins can be synthesized. In 2006, French researchers came across a puzzling example of RNA-mediated inheritance in mice. Mice with a loss-of-function mutation in the gene Kit have white tails. Offspring of these mutants can have white tails despite having only normal Kit genes. The research team traced this effect back to mutated Kit RNA. While RNA is common as genetic storage material in viruses, in mammals in particular RNA inheritance has been observed very rarely.

Functional Structure of a Gene

The vast majority of living organisms encode their genes in long strands of DNA. DNA (deoxyribonucleic acid) consists of a chain made from four types of nucleotide subunits, each composed of: a five-carbon sugar (2'-deoxyribose), a phosphate group, and one of the four bases adenine, cytosine, guanine, and thymine. The most common form of DNA in a cell is in a double helix structure, in which two individual DNA strands twist around each other in a right-handed spiral. In this structure, the base pairing rules specify that guanine pairs with cytosine and adenine pairs with thymine. The base pairing between guanine and cytosine forms three hydrogen bonds, whereas the base pairing between adenine and thymine forms two hydrogen bonds. The two strands in a double helix must therefore be *complementary*, that is, their bases must align such that the adenines of one strand are paired with the thymines of the other strand, and so on.

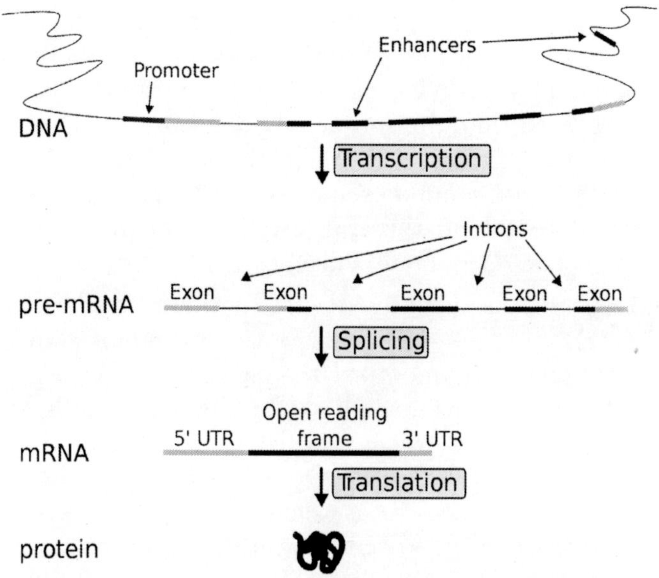

Figure : *Diagram of the "typical" eukaryotic protein-coding gene. Promoters and enhancers determine what portions of the DNA will be transcribed into the precursor mRNA (pre-mRNA). The pre-mRNA is then spliced into messenger RNA (mRNA) which is later translated into protein.*

Due to the chemical composition of the pentose residues of the bases, DNA strands have directionality. One end of a DNA polymer contains an exposed hydroxyl group on the deoxyribose; this is known as the 3' end of the molecule. The other end contains an exposed phosphate group; this is the 5' end. The directionality of DNA is vitally important to many cellular processes, since double helices are necessarily directional (a strand running 5'-3' pairs with a complementary strand running 3'-5'), and processes such as DNA replication occur in only one direction.

All nucleic acid synthesis in a cell occurs in the 5'-3' direction, because new monomers are added via a dehydration reaction that uses the exposed 3' hydroxyl as a nucleophile. The expression of genes encoded in DNA begins by transcribing the gene into RNA, a second type of nucleic acid that is very similar to DNA, but whose monomers contain the sugar ribose rather than deoxyribose. RNA also contains the base uracil in place of thymine. RNA molecules are less stable than DNA and are typically single-stranded. Genes that encode proteins are composed of a series of three-nucleotide sequences called codons, which serve as the *words* in the genetic *language*. The genetic code specifies the correspondence during protein translation between codons and amino acids. The genetic code is nearly the same for all known organisms.

All genes have regulatory regions in addition to regions that explicitly code for a protein or RNA product. A regulatory region shared by almost all genes is known as the promoter, which provides a position that is recognised by the transcription machinery when a gene is about to be transcribed and expressed. A gene can have more than one promoter, resulting in RNAs that differ in how far they extend in the 5' end. Although promoter regions have a consensus sequence that is the most common sequence at this position, some genes have "strong" promoters that bind the transcription machinery well, and others have "weak" promoters that bind poorly.

These weak promoters usually permit a lower rate of transcription than the strong promoters, because the transcription machinery binds to them and initiates transcription less frequently. Other possible regulatory regions include enhancers, which can compensate for a weak promoter. Most regulatory regions are "upstream"—that is, before or toward the 5' end of the transcription initiation site. Eukaryotic promoter regions are much more complex and difficult to identify than prokaryotic promoters. Many prokaryotic genes are organised into operons, or groups of genes whose products have related functions and which are transcribed as a unit. By contrast, eukaryotic genes are transcribed only one at a time, but may include long stretches of DNA called introns which are transcribed but never translated into protein (they are spliced out before translation). Splicing can also occur in prokaryotic genes, but is less common than in eukaryotes.

Chromosomes

The total complement of genes in an organism or cell is known as its genome, which may be stored on one or more chromosomes; the region of the chromosome at which a particular gene is located is called its locus. A chromosome consists of a single, very long DNA helix on which thousands of genes are encoded. Prokaryotes—bacteria and archaea—typically store their genomes on a single large, circular chromosome, sometimes supplemented by additional small circles of DNA called plasmids, which usually encode only a few genes and are easily transferable between individuals. For example, the genes for antibiotic resistance are usually encoded on bacterial plasmids and can be passed between individual cells, even those of different species, via horizontal gene transfer. Although some simple eukaryotes also possess plasmids with small numbers of genes, the majority of eukaryotic genes are stored on multiple linear chromosomes, which are packed within the nucleus in complex with storage proteins called histones. The manner in which DNA is stored on the histone, as well as chemical modifications of the histone itself, are regulatory mechanisms governing whether a particular region of DNA is accessible for gene expression.

The ends of eukaryotic chromosomes are capped by long stretches of repetitive sequences called telomeres, which do not code for any gene product but are present to prevent degradation of coding and regulatory regions during DNA replication. The length of the telomeres tends to decrease each time the genome is replicated in preparation for cell division; the loss of telomeres has been proposed as an explanation for cellular senescence, or the loss of the ability to divide, and by extension for the aging process in organisms.

Whereas the chromosomes of prokaryotes are relatively gene-dense, those of eukaryotes often contain so-called "junk DNA", or regions of DNA that serve no obvious function. Simple single-celled eukaryotes have relatively small amounts of such DNA, whereas the genomes of complex multicellular organisms, including humans, contain an absolute majority of DNA without an identified function. However it now appears that, although protein-coding DNA makes up barely 2% of the human genome, about 80% of the bases in the genome may be expressed, so the term "junk DNA" may be a misnomer.

Gene Expression

In all organisms, there are two major steps separating a protein-coding gene from its protein: First, the DNA on which the gene resides must be *transcribed* from DNA to messenger RNA (mRNA); and, second, it must be *translated* from mRNA to protein. RNA-coding genes must still go through the first step, but are not translated into protein. The process of producing a biologically functional molecule of either RNA or protein is called gene expression, and the resulting molecule itself is called a gene product.

Genetic Code

The genetic code is the set of rules by which a gene is translated into a functional protein. Each gene consists of a specific sequence of nucleotides encoded in a DNA (or sometimes RNA in some viruses) strand; a correspondence between nucleotides, the basic building blocks of genetic material, and amino acids, the basic building blocks of proteins, must be established for genes to be successfully translated into functional proteins.

Sets of three nucleotides, known as codons, each correspond to a specific amino acid or to a signal; three codons are known as "stop codons" and, instead of specifying a new amino acid, alert the translation machinery that the end of the gene has been reached.

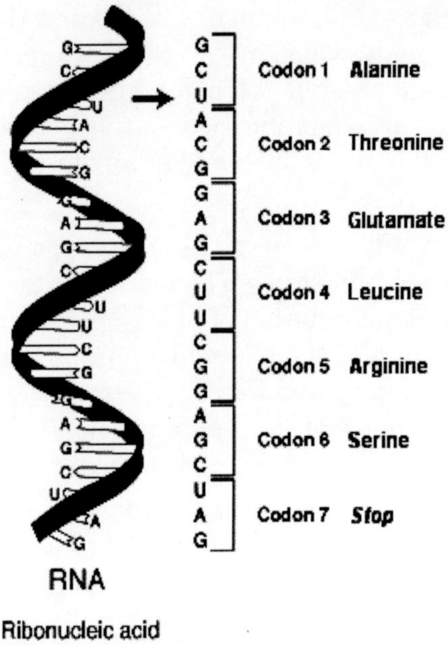

Figure : *Schematic diagram of a single-stranded RNA molecule illustrating the position of three-base codons.*

There are 64 possible codons (four possible nucleotides at each of three positions, hence 4^3 possible codons) and only 20 standard amino acids; hence the code is redundant and multiple codons can specify the same amino acid. The correspondence between codons and amino acids is nearly universal among all known living organisms.

Transcription

The process of genetic transcription produces a single-stranded RNA molecule known as messenger RNA, whose nucleotide sequence is complementary to the DNA from which it was transcribed. The DNA strand whose sequence matches that of the RNA is known as the coding strand and the strand from which the RNA was synthesized is the template strand. Transcription is performed by an enzyme called an RNA polymerase, which reads the template strand in the 3' to 5' direction and synthesizes the RNA from 5' to 3'. To initiate transcription, the polymerase first recognises and binds a promoter region of the gene. Thus a major mechanism of gene regulation is the blocking or sequestering of the promoter region, either by tight binding by repressor molecules that physically block the polymerase, or by organising the DNA so that the promoter region is not accessible.

In prokaryotes, transcription occurs in the cytoplasm; for very long transcripts, translation may begin at the 5' end of the RNA while the 3' end is still being transcribed. In eukaryotes, transcription necessarily occurs in the nucleus, where the cell's DNA is sequestered; the RNA molecule produced by the polymerase is known as the primary transcript and must undergo post-transcriptional modifications before being exported to the cytoplasm for translation. The splicing of introns present within the transcribed region is a modification unique to eukaryotes; alternative splicing mechanisms can result in mature transcripts from the same gene having different sequences and thus coding for different proteins. This is a major form of regulation in eukaryotic cells.

Translation

Translation is the process by which a mature mRNA molecule is used as a template for synthesizing a new protein. Translation is carried out by ribosomes, large complexes of RNA and protein responsible for carrying out the chemical reactions to add new amino acids to a growing polypeptide chain by the formation of peptide bonds.

The genetic code is read three nucleotides at a time, in units called codons, via interactions with specialised RNA molecules called transfer RNA (tRNA). Each tRNA has three unpaired bases known as the anticodon that are complementary to the codon it reads; the tRNA is also covalently attached to the amino acid specified by the complementary codon. When the tRNA binds to its complementary codon in an mRNA strand, the ribosome ligates its amino acid cargo to the new polypeptide chain, which is synthesized from amino terminus to carboxyl terminus. During and after its synthesis, the new protein must fold to its active three-dimensional structure before it can carry out its cellular function.

DNA Replication and Inheritance

The growth, development, and reproduction of organisms relies on cell division, or the process by which a single cell divides into two usually identical daughter cells. This requires first making a duplicate copy of every gene in the genome in a process called DNA replication. The copies are made by specialised enzymes known as DNA polymerases, which "read" one strand of the double-helical DNA, known as the template strand, and synthesize a new complementary strand. Because the DNA double helix is held together by base pairing, the sequence of one strand completely specifies the sequence of its complement; hence only one strand needs to be read by the enzyme to produce a faithful copy.

The process of DNA replication is semiconservative; that is, the copy of the genome inherited by each daughter cell contains one original and one newly synthesized strand of DNA. After DNA replication is complete, the cell must physically separate the two copies of the genome and divide into two distinct membrane-bound cells. In prokaryotes - bacteria and archaea - this usually occurs via a relatively simple process called binary fission, in which each circular genome attaches to the cell membrane and is separated into the daughter cells as the membrane invaginates to split the cytoplasm into two membrane-bound portions. Binary fission is extremely fast compared to the rates of cell division in eukaryotes. Eukaryotic cell division is a more complex process known as the cell cycle; DNA replication occurs during a phase of this cycle known as S phase, whereas the process of segregating chromosomes and splitting the cytoplasm occurs during M phase. In many single-celled eukaryotes such as yeast, reproduction by budding is common, which results in asymmetrical portions of cytoplasm in the two daughter cells.

Molecular Inheritance

The duplication and transmission of genetic material from one generation of cells to the next is the basis for molecular inheritance, and the link between the classical and molecular pictures of genes. Organisms inherit the characteristics of their parents because the cells of the offspring contain copies of the genes in their parents' cells. In asexually reproducing organisms, the offspring will be a genetic copy or clone of the parent organism. In sexually reproducing organisms, a specialised form of cell division called meiosis produces cells called gametes or germ cells that are haploid, or contain only one copy of each gene. The gametes produced by females are called eggs or ova, and those produced by males are called sperm. Two gametes fuse to form a fertilized egg, a single cell that once again has a diploid number of genes—each with one copy from the mother and one copy from the father. During the process of meiotic cell division, an event called genetic recombination or *crossing-over* can sometimes occur, in which a length of DNA on one chromatid is swapped with a length of DNA on the corresponding sister chromatid. This has no effect if the alleles on the chromatids are the same, but results in reassortment of otherwise linked alleles if they are different. The Mendelian principle of independent assortment asserts that each of a parent's two genes for each trait will sort independently into gametes; which allele an organism inherits for one trait is unrelated to which allele it inherits for another trait. This is in fact only true for genes that do not reside on the same chromosome, or are located very far from one another on the same chromosome. The closer two genes lie on the same chromosome, the more closely they will be associated in gametes and the more often they will appear together; genes that are very close are essentially never separated because it is extremely unlikely that a crossover point will occur between them. This is known as genetic linkage.

History

The notion of a gene is evolving with the science of genetics, which began when Gregor Mendel noticed that biological variations are inherited from parent organisms as specific, discrete traits. The biological entity responsible for defining traits was later termed a *gene*, but the biological basis for inheritance remained unknown until DNA was identified as the genetic material in the 1940s. Prior to Mendel's work, the dominant theory of heredity was one of blending inheritance, which proposes that the traits of the parents blend or mix in a smooth, continuous gradient in the offspring. Although Mendel's work was largely unrecognised after its first publication in 1866, it was rediscovered in 1900 by three European scientists, Hugo de Vries, Carl Correns, and Erich von Tschermak, who had reached similar conclusions from their own research. However, these scientists were not yet aware of the identity of the 'discrete units' on which genetic material resides.

The existence of genes was first suggested by Gregor Mendel (1822–1884), who, in the 1860s, studied inheritance in peaplants (*Pisum sativum*) and hypothesized a factor that conveys traits from parent to offspring. He spent over 10 years of his life on one experiment. Although he did not use the term *gene*, he explained his results in terms of inherited characteristics. Mendel was also the first to hypothesize independent assortment, the distinction between dominant and recessive traits, the distinction between a heterozygote and homozygote, and the difference between what would later be described as genotype (the genetic material of an organism) and phenotype (the visible traits of that organism).

Charles Darwin used the term Gemmule to describe a microscopic unit of inheritance, and what would later become known as Chromosomes had been observed separating out during cell division by Wilhelm Hofmeister as early as 1848. The idea that chromosomes are the carriers of inheritance was expressed in 1883 by Wilhelm Roux. Darwin also coined the word *pangenesis* by (1868). The word pangenesis is made from the Greek words *pan* (a prefix meaning "whole", "encompassing") and *genesis* ("birth") or *genos* ("origin").

Mendel's concept was given a name by Hugo de Vries in 1889, in his book *Intracellular Pangenesis*; although probably unaware of Mendel's work at the time, he coined the term "pangen" for "the smallest particle [representing] one hereditary characteristic". Danish botanist Wilhelm Johannsen coined the word "gene" ("gen" in Danish and German) in 1909 to describe the fundamental physical and functional units of heredity, while the related word genetics was first used by William Bateson in 1905. He derived the word from de Vries' "pangen". In the early 1990s, Mendel's work received renewed attention from scientists. In 1910, Thomas Hunt Morgan showed that genes reside on specific chromosomes. He later showed that genes occupy specific locations on the chromosome. With this knowledge, Morgan and his students began the first chromosomal map of the fruit fly *Drosophila*. In 1928, Frederick Griffith showed that genes could be transferred. In what is now known as Griffith's experiment, injections into a mouse of a deadly strain of bacteria that had been heat-killed transferred genetic information to a safe strain of the same bacteria, killing the mouse. A series of subsequent discoveries led to the realisation decades later that chromosomes within cells are the carriers of genetic material, and that they are made of DNA (deoxyribonucleic acid), a polymeric molecule found in all cells on which the 'discrete units' of Mendelian inheritance are encoded. In 1941, George Wells Beadle and Edward Lawrie Tatum showed that mutations in genes caused errors in specific steps in metabolic pathways. This showed that specific genes code for specific proteins, leading to the "one gene, one enzyme" hypothesis.

Oswald Avery, Colin Munro MacLeod, and Maclyn McCarty showed in 1944 that DNA holds the gene's information. In 1953, James D. Watson and Francis Crick demonstrated the molecular structure of DNA. Together, these discoveries established the central dogma of molecular biology, which states that proteins are translated from RNA which is transcribed from DNA. This dogma has since been shown to have exceptions, such as reverse transcription in retroviruses. In 1972, Walter Fiers and his team at the Laboratory of Molecular Biology of the University of Ghent (Ghent, Belgium) were the first to determine the sequence of a gene: the gene for Bacteriophage MS2 coat protein. Richard J. Roberts and Phillip Sharp discovered in 1977 that genes can be split into segments.

This led to the idea that one gene can make several proteins. Recently (as of 2003–2006), biological results let the notion of gene appear more slippery. In particular, genes do not seem to sit side by side on DNA like discrete beads. Instead, regions of the DNA producing distinct proteins may overlap, so that the idea emerges that "genes are one long continuum". It was first hypothesized in 1986 by Walter Gilbert that neither DNA nor protein would be required in such a primitive system as that of a very early stage of the earth if RNA could perform as simply a catalyst and genetic information storage processor. The modern study of genetics at the level of DNA is known as molecular genetics and the synthesis of molecular genetics with traditional Darwinian evolution is known as the modern evolutionary synthesis.

Mendelian Inheritance and Classical Genetics

According to the theory of Mendelian inheritance, variations in phenotype—the observable physical and behavioural characteristics of an organism—are due to variations in genotype, or the organism's particular set of genes, each of which specifies a particular trait. Different forms of a gene, which may give rise to different phenotypes, are known as alleles. Organisms such as the pea plants Mendel worked on, along with many plants and animals, have two alleles for each trait, one inherited from each parent. Alleles may be dominant or recessive; dominant alleles give rise to their corresponding phenotypes when paired with any other allele for the same trait, whereas recessive alleles give rise to their corresponding phenotype only when paired with another copy of the same allele. For example, if the allele specifying tall stems in pea plants is dominant over the allele specifying short stems, then pea plants that inherit one tall allele from one parent and one short allele from the other parent will also have tall stems. Mendel's work found that alleles assort independently in the production of gametes, or germ cells, ensuring variation in the next generation.

Mutation

DNA replication is for the most part extremely accurate, with an error rate per site of around 10^{-6} to 10^{-10} in eukaryotes. Rare, spontaneous alterations in the base sequence of a particular gene arise from a number of sources, such as errors in DNA replication and the aftermath of DNA damage. These errors are called mutations. The cell contains many DNA repair mechanisms for preventing mutations and maintaining the integrity of the genome; however, in some cases—such as breaks in both DNA strands of a chromosome—repairing the physical damage to the molecule is a higher priority than producing an exact copy. Due to the degeneracy of the genetic code, some mutations in protein-coding genes are *silent*, or produce no change in the amino acid sequence of the protein for which they code; for example, the codons UCU and UUC both code for serine, so the U"!C mutation has no effect on the protein. Mutations that do have phenotypic effects are most often neutral or deleterious to the organism, but sometimes they confer benefits to the organism's fitness.

Mutations propagated to the next generation lead to variations within a species' population. Variants of a single gene are known as alleles, and differences in alleles may give rise to differences in traits. Although it is rare for the variants in a single gene to have clearly distinguishable phenotypic effects, certain well-defined traits are in fact controlled by single genetic loci. A gene's most common allele is called the wild type allele, and rare alleles are called mutants. However, this does not imply that the wild-type allele is the ancestor from which the mutants are descended.

Chromosomal Organisation

The total complement of genes in an organism or cell is known as its genome. In prokaryotes, the vast majority of genes are located on a single chromosome of circular DNA, while eukaryotes usually possess multiple individual linear DNA helices packed into dense DNA-protein complexes called chromosomes. Genes that appear together on one chromosome of one species may appear on separate chromosomes in another species. Many species carry more than one copy of their genome within each of their somatic cells. Cells or organisms with only one copy of each chromosome are called haploid; those with two copies are called diploid; and those with more than two copies are called polyploid. The copies of genes on the chromosomes are not necessarily identical. In sexually reproducing organisms, one copy is normally inherited from each parent.

Number of Genes

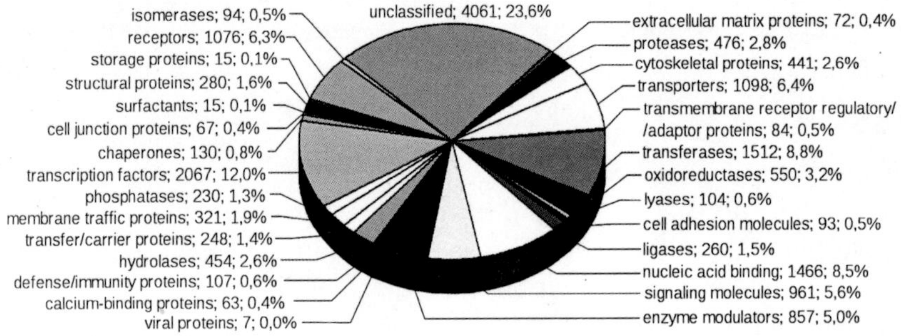

Figure : *The human genome, categorized by function of each gene product, given both as number of genes and as percentage of all genes.*

Early estimates of the number of human genes that used expressed sequence tag data put it at 50 000–100 000. Following the sequencing of the human genome and other genomes, it has been found that rather few genes (~20 000 in human, mouse and fly, ~13 000 in roundworm, >46 000 in rice) encode all the proteins in an organism. These protein-coding sequences make up 1–2% of the human genome. A large part of the genome is transcribed however, to introns, retrotransposons and seemingly a large array of noncoding RNAs. Total number of proteins (the Earth's proteome) is estimated to be 5 million sequences.

Genetic and Genomic Nomenclature

Gene nomenclature has been established by the HUGO Gene Nomenclature Committee (HGNC) for each known human gene in the form of an approved gene name and symbol (short-form abbreviation). All approved symbols are stored in the HGNC Database. Each symbol is unique and each gene is only given one approved gene symbol. This also facilitates electronic data retrieval from publications. In preference each symbol maintains parallel construction in different members of a gene family and can be used in other species, especially the mouse.

Evolutionary Concept of a Gene

George C. Williams first explicitly advocated the gene-centric view of evolution in his 1966 book *Adaptation and Natural Selection*. He proposed an evolutionary concept of gene to be used when we are talking about natural selection favouring some genes. The definition is: "that which segregates and recombines with appreciable frequency." According to this definition, even an asexual genome could be considered a gene, insofar that it have an appreciable permanency through many generations.

The difference is: the molecular gene *transcribes* as a unit, and the evolutionary gene *inherits* as a unit. Richard Dawkins' books *The Selfish Gene* (1976) and *The Extended Phenotype* (1982) defended the idea that the gene is the only replicator in living systems. This means that only genes transmit their structure largely intact and are potentially immortal in the form of copies. So, genes should be the unit of selection.

In *The Selfish Gene* Dawkins attempts to redefine the word 'gene' to mean "an inheritable unit" instead of the generally accepted definition of "a section of DNA coding for a particular protein". In *River Out of Eden*, Dawkins further refined the idea of gene-centric selection by describing life as a river of compatible genes flowing through geological time. Scoop up a bucket of genes from the river of genes, and we have an organism serving as temporary bodies or survival machines. A river of genes may fork into two branches representing two non-interbreeding species as a result of geographical separation.

Gene Targeting and Implications

Gene targeting is commonly referred to techniques for altering or disrupting mouse genes and provides the mouse models for studying the roles of individual genes in embryonic development, human disorders, aging and diseases.

The mouse models, where one or more of its genes are deactivated or made inoperable, are called knockout mice. Since the first reports in which homologous recombination in embryonic stem cells was used to generate gene-targeted mice, gene targeting has proven to be a powerful means of precisely manipulating the mammalian genome, producing at least ten thousand mutant mouse strains and it is now possible to introduce mutations that can be activated at specific time points, or in specific cells or organs, both during development and in the adult animal.

Gene targeting strategies have been expanded to all kinds of modifications, including point mutations, isoform deletions, mutant allele correction, large pieces of chromosomal DNA insertion and deletion, tissue specific disruption combined with spatial and temporal regulation and so on. It is predicted that the ability to generate mouse models with predictable phenotypes will have a major impact on studies of all phases of development, immunology, neurobiology, oncology, physiology, metabolism, and human diseases. Gene targeting is also in theory applicable to species from which totipotent embryonic stem cells can be established, and therefore may offer a potential to the improvement of domestic animals and plants.

Changing Concept

The concept of the gene has changed considerably. From the original definition of a "unit of inheritance", the term evolved to mean a DNA-based unit that can exert its effects on the organism through RNA or protein products. It was also previously believed that one gene makes one protein; this concept was overthrown by the discovery of alternative splicing and trans-splicing. The definition of a gene is still changing. The first cases of RNA-based inheritance have been discovered in mammals. Evidence is also accumulating that the control regions of a gene do not necessarily have to be close to the coding sequence on the linear molecule or even on the same chromosome. Spilianakis and colleagues discovered that the promoter region of the interferon-gamma gene on chromosome 10 and the regulatory regions of the T(H)2 cytokine locus on chromosome 11 come into close proximity in the nucleus possibly to be jointly regulated.

The concept that genes are clearly delimited is also being eroded. There is evidence for fused proteins stemming from two adjacent genes that can produce two separate protein products. While it is not clear whether these fusion proteins are functional, the phenomenon is more frequent than previously thought. Even more ground-breaking than the discovery of fused genes is the observation that some proteins can be composed of exons from far away regions and even different chromosomes. This new data has led to an updated, and probably tentative, definition of a gene as "a union of genomic sequences encoding a coherent set of potentially overlapping functional products". This new definition categorises genes by functional products, whether they be proteins or RNA, rather than specific DNA loci; all regulatory elements of DNA are therefore classified as *gene-associated* regions.

DNA

Deoxyribonucleic acid is a nucleic acid that contains the genetic instructions used in the development and functioning of all known living organisms (with the exception of RNA viruses). The DNA segments that carry this genetic information are called genes, but other DNA sequences have structural purposes, or are involved in regulating the use of this genetic information. Along with RNA and proteins, DNA is one of the three major macromolecules that are essential for all known forms of life. DNA consists of two long polymers of simple units called nucleotides, with backbones made of sugars and phosphate groups joined by ester bonds. These two strands run in opposite directions to each other and are therefore anti-parallel. Attached to each sugar is one of four types of molecules called nucleobases (informally, *bases*). It is the sequence of these four nucleobases along the backbone that encodes information. This information is read using the genetic code, which specifies the sequence of the amino acids within proteins. The code is read by copying stretches of DNA into the related nucleic acid RNA in a process called transcription. Within cells DNA is organised into long structures called chromosomes. During cell division these chromosomes are duplicated in the process of DNA replication, providing each cell its own complete set of chromosomes. Eukaryotic organisms (animals, plants, fungi, and protists) store most of their DNA inside the cell nucleus and some of their DNA in organelles, such as mitochondria or chloroplasts. In contrast, prokaryotes (bacteria and archaea) store their DNA only in the cytoplasm. Within the chromosomes, chromatin proteins such as histones compact and organise DNA. These compact structures guide the interactions between DNA and other proteins, helping control which parts of the DNA are transcribed.

Properties

Chemical structure of DNA. Hydrogen bonds shown as dotted lines. DNA is a long polymer made from repeating units called nucleotides. As first discovered by James D. Watson and Francis Crick, the structure of DNA of all species comprises two helical chains each coiled round the same axis, and each with a pitch of 34 Ångströms (3.4 nanometres) and a radius of 10 Ångströms (1.0 nanometres). According to another study, when measured in a particular solution, the DNA chain measured 22 to 26 Ångströms wide (2.2 to 2.6 nanometres), and one nucleotide unit measured 3.3 Å (0.33 nm) long. Although each individual repeating unit is very small, DNA polymers can be very large molecules containing millions of nucleotides. For instance, the largest human chromosome, chromosome number 1, is approximately 220 million base pairs long. In living organisms DNA does not usually exist as a single molecule, but instead as a pair of molecules that are held tightly together. These two long strands entwine like vines, in the shape of a double helix. The nucleotide repeats contain both the segment of the backbone of the molecule, which holds the chain together, and a nucleobase, which interacts with the other DNA strand in the helix. A nucleobase linked to a sugar is called a nucleoside and a base linked to a sugar and one or more phosphate groups is called a nucleotide. Polymers comprising multiple linked nucleotides (as in DNA) are called a polynucleotide.

The backbone of the DNA strand is made from alternating phosphate and sugar residues. The sugar in DNA is 2-deoxyribose, which is a pentose (five-carbon) sugar. The sugars are joined together by phosphate groups that form phosphodiester bonds between the third and fifth carbon atoms of adjacent sugar rings. These asymmetric bonds mean a strand of DNA has a direction. In a double helix the direction of the nucleotides in one strand is opposite to their direction in the other strand: the strands are *antiparallel*. The asymmetric ends of DNA strands are called the 52 (*five prime*) and 32 (*three prime*) ends, with the 5' end having a terminal phosphate group and the 3' end a terminal hydroxyl group. One major difference between DNA and RNA is the sugar, with the 2-deoxyribose in DNA being replaced by the alternative pentose sugar ribose in RNA. A section of DNA. The bases lie horizontally between the two spiraling strands. Animated version at File:DNA orbit animated.gif.

The DNA double helix is stabilized primarily by two forces: hydrogen bonds between nucleotides and base-stacking interactions among the aromatic nucleobases. In the aqueous environment of the cell, the conjugated ∂ bonds of nucleotide bases align perpendicular to the axis of the DNA molecule, minimising their interaction with the solvation shell and therefore, the Gibbs free energy. The four bases found in DNA are adenine (abbreviated A), cytosine (C), guanine (G) and thymine (T). These four bases are attached to the sugar/phosphate to form the complete nucleotide, as shown for adenosine monophosphate. The nucleobases are classified into two types: the purines, A and G, being fused five- and six-membered heterocyclic compounds, and the pyrimidines, the six-membered rings C and T. A fifth pyrimidine nucleobase, uracil (U), usually takes the place of thymine in RNA and differs from thymine by lacking a methyl group on its ring. Uracil is not usually found in DNA, occurring only as a breakdown product of cytosine. In addition to RNA and DNA a large number of artificial nucleic acid analogues have also been created to study the proprieties of nucleic acids, or for use in biotechnology.

Figure : *Major and minor grooves of DNA. Minor groove is a binding site for the dye Hoechst 33258.*

Grooves

Twin helical strands form the DNA backbone. Another double helix may be found by tracing the spaces, or grooves, between the strands. These voids are adjacent to the base pairs and may provide a binding site. As the strands are not directly opposite each other, the grooves are unequally sized. One groove, the major groove, is 22 Å wide and the other, the minor groove, is 12 Å wide. The narrowness of the minor groove means that the edges of the bases are more accessible in the major groove. As a result, proteins like transcription factors that can bind to specific sequences in double-stranded DNA usually make contacts to the sides of the bases exposed in the major groove. This situation varies in unusual conformations of DNA within the cell, but the major and minor grooves are always named to reflect the differences in size that would be seen if the DNA is twisted back into the ordinary B form.

Base Pairing

In a DNA double helix, each type of nucleobase on one strand normally interacts with just one type of nucleobase on the other strand. This is called complementary base pairing. Here, purines form hydrogen bonds to pyrimidines, with A bonding only to T, and C bonding only to G. This arrangement of two nucleotides binding together across the double helix is called a base pair. As hydrogen bonds are not covalent, they can be broken and rejoined relatively easily. The two strands of DNA in a double helix can therefore be pulled apart like a zipper, either by a mechanical force or high temperature. As a result of this complementarity, all the information in the double-stranded sequence of a DNA helix is duplicated on each strand, which is vital in DNA replication. Indeed, this reversible and specific interaction between complementary base pairs is critical for all the functions of DNA in living organisms.

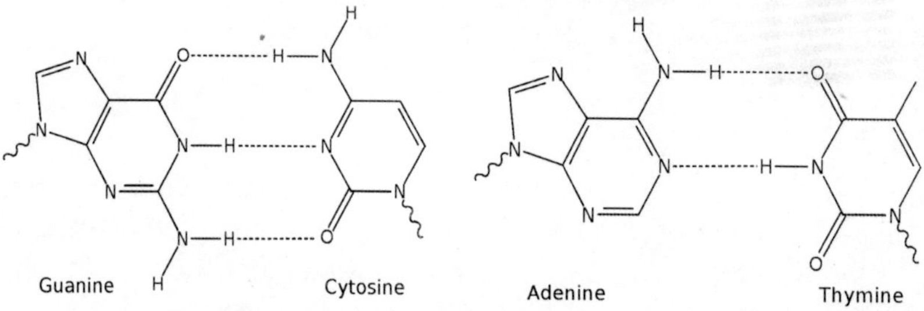

Figure : *Top, a GC base pair with three hydrogen bonds. Bottom, an AT base pair with two hydrogen bonds. Non-covalent hydrogen bonds between the pairs are shown as dashed lines.*

The two types of base pairs form different numbers of hydrogen bonds, AT forming two hydrogen bonds, and GC forming three hydrogen bonds. DNA with high GC-content is more stable than DNA with low GC-content. Although it is often stated that this is due to the added stability of an additional hydrogen bond, this is incorrect. DNA with high GC-content is more stable due to intra-strand base stacking interactions. As noted above, most DNA molecules are actually two polymer strands, bound together in a helical fashion by noncovalent bonds; this double stranded structure (dsDNA) is maintained largely by the intrastrand base stacking interactions, which are strongest for

G,C stacks. The two strands can come apart – a process known as melting – to form two ss DNA molecules. Melting occurs when conditions favour ssDNA; such conditions are high temperature, low salt and high pH (low pH also melts DNA, but since DNA is unstable due to acid depurination, low pH is rarely used). The stability of the dsDNA form depends not only on the GC-content (% G,C basepairs) but also on sequence (since stacking is sequence specific) and also length (longer molecules are more stable).

The stability can be measured in various ways; a common way is the "melting temperature", which is the temperature at which 50% of the ds molecules are converted to ss molecules; melting temperature is dependent on ionic strength and the concentration of DNA. As a result, it is both the percentage of GC base pairs and the overall length of a DNA double helix that determine the strength of the association between the two strands of DNA. Long DNA helices with a high GC-content have stronger-interacting strands, while short helices with high AT content have weaker-interacting strands. In biology, parts of the DNA double helix that need to separate easily, such as the TATAAT Pribnow box in some promoters, tend to have a high AT content, making the strands easier to pull apart. In the laboratory, the strength of this interaction can be measured by finding the temperature required to break the hydrogen bonds, their melting temperature (also called T_m value). When all the base pairs in a DNA double helix melt, the strands separate and exist in solution as two entirely independent molecules. These single-stranded DNA molecules (*ssDNA*) have no single common shape, but some conformations are more stable than others.

Sense and Antisense

A DNA sequence is called "sense" if its sequence is the same as that of a messenger RNA copy that is translated into protein. The sequence on the opposite strand is called the "antisense" sequence. Both sense and antisense sequences can exist on different parts of the same strand of DNA (i.e. both strands contain both sense and antisense sequences). In both prokaryotes and eukaryotes, antisense RNA sequences are produced, but the functions of these RNAs are not entirely clear. One proposal is that antisense RNAs are involved in regulating gene expression through RNA-RNA base pairing. A few DNA sequences in prokaryotes and eukaryotes, and more in plasmids and viruses, blur the distinction between sense and antisense strands by having overlapping genes. In these cases, some DNA sequences do double duty, encoding one protein when read along one strand, and a second protein when read in the opposite direction along the other strand. In bacteria, this overlap may be involved in the regulation of gene transcription, while in viruses, overlapping genes increase the amount of information that can be encoded within the small viral genome.

Supercoiling

DNA can be twisted like a rope in a process called DNA supercoiling. With DNA in its "relaxed" state, a strand usually circles the axis of the double helix once every 10.4 base pairs, but if the DNA is twisted the strands become more tightly or more loosely wound. If the DNA is twisted in the direction of the helix, this is positive supercoiling, and the bases are held more tightly together. If they are twisted in the opposite direction, this is negative supercoiling, and the bases come apart more easily. In nature, most DNA has slight negative supercoiling that is introduced by enzymes called to poisomerases. These enzymes are also needed to relieve the twisting stresses introduced into DNA strands during processes such as transcription and DNA replication.

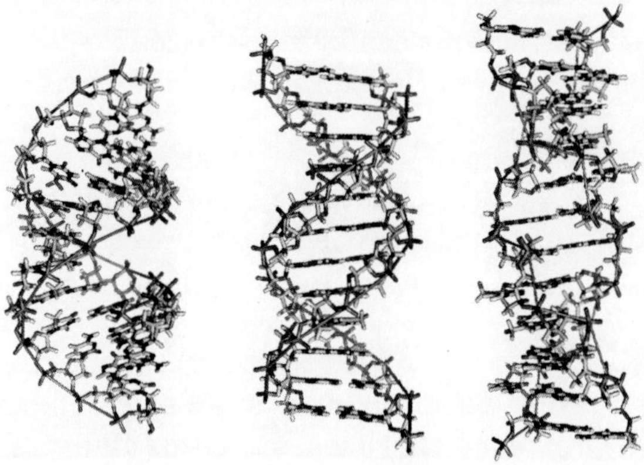

***Figure :** From left to right, the structures of A, B and Z DNA*

Alternate DNA Structures

DNA exists in many possible conformations that include A-DNA, B-DNA, and Z-DNA forms, although, only B-DNA and Z-DNA have been directly observed in functional organisms. The conformation that DNA adopts depends on the hydration level, DNA sequence, the amount and direction of supercoiling, chemical modifications of the bases, the type and concentration of metal ions, as well as the presence of polyamines in solution.

The first published reports of A-DNA X-ray diffraction patterns— and also B-DNA used analyses based on Patterson transforms that provided only a limited amount of structural information for oriented fibres of DNA.

An alternate analysis was then proposed by Wilkins *et al.*, in 1953, for the *in vivo* B-DNA X-ray diffraction/scattering patterns of highly hydrated DNA fibres in terms of squares of Bessel functions. In the same journal, James D. Watson and Francis Crick presented their molecular modelling analysis of the DNA X-ray diffraction patterns to suggest that the structure was a double-helix. Although the 'B-DNA form' is most common under the conditions found in cells, it is not a well-defined conformation but a family of related DNA conformations that occur at the high hydration levels present in living cells. Their corresponding X-ray diffraction and scattering patterns are characteristic of molecular paracrystals with a significant degree of disorder.

Compared to B-DNA, the A-DNA form is a wider right-handed spiral, with a shallow, wide minor groove and a narrower, deeper major groove. The A form occurs under non-physiological conditions in partially dehydrated samples of DNA, while in the cell it may be produced in hybrid pairings of DNA and RNA strands, as well as in enzyme-DNA complexes. Segments of DNA where the bases have been chemically modified by methylation may undergo a larger change in conformation and adopt the Z form. Here, the strands turn about the helical axis in a left-handed spiral, the opposite of the more common B form. These unusual structures can be recognised by specific Z-DNA binding proteins and may be involved in the regulation of transcription.

Alternate DNA Chemistry

For a number of years exobiologists have proposed the existence of a shadow biosphere, a postulated microbial biosphere of Earth that uses radically different biochemical and molecular processes than currently known life. One of the proposals was the existence of lifeforms that use arsenic instead of phosphorus in DNA. A December 2010 NASA press conference stated that the bacterium GFAJ-1, which has evolved in an arsenic-rich environment, is the first terrestrial lifeform found which may have this ability. The bacterium was found in Mono Lake, east of Yosemite National Park. GFAJ-1 is a rod-shaped extremophile bacterium in the family Halomonadaceae that, when starved of phosphorus, may be capable of incorporating the usually poisonous element arsenic in its DNA.

This discovery may lend weight to the long-standing idea that extraterrestrial life could have a different chemical makeup from life on Earth. The research was carried out by a team led by Felisa Wolfe-Simon, a geomicrobiologist and geobiochemist, a Postdoctoral Fellow of the NASA Astrobiology Institute with Arizona State University. This finding has, however, faced strong criticism from the scientific community; scientists have argued that there is no evidence that arsenic is actually incorporated into biomolecules. Independent conformation of this finding has also not yet been possible.

Quadruplex Structures

At the ends of the linear chromosomes are specialised regions of DNA called telomeres. The main function of these regions is to allow the cell to replicate chromosome ends using the enzyme telomerase, as the enzymes that normally replicate DNA cannot copy the extreme 32 ends of chromosomes. These specialised chromosome caps also help protect the DNA ends, and stop the DNA repair systems in the cell from treating them as damage to be corrected. In human cells, telomeres are usually lengths of single-stranded DNA containing several thousand repeats of a simple TTAGGG sequence.

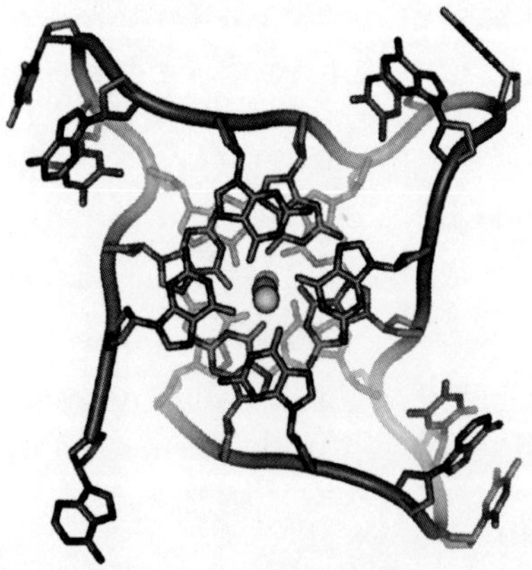

Figure : *DNA quadruplex formed by telomere repeats. The looped conformation of the DNA backbone is very different from the typical DNA helix.*

These guanine-rich sequences may stabilize chromosome ends by forming structures of stacked sets of four-base units, rather than the usual base pairs found in other DNA molecules. Here, four guanine bases form a flat plate and these flat four-base units then stack on top of each other, to form a stable *G-quadruplex* structure. These structures are stabilized by hydrogen bonding between the edges of the bases and chelation of a metal ion in the centre of each four-base unit. Other structures can also be formed, with the central set of four bases coming from either a single strand folded around the bases, or several different parallel strands, each contributing one base to the central structure. In addition to these stacked structures, telomeres also form large loop structures called telomere loops, or T-loops. Here, the single-stranded DNA curls around in a long circle stabilized by telomere-binding proteins. At the very end of the T-loop, the single-stranded telomere DNA is held onto a region of double-stranded DNA by the telomere strand disrupting the double-helical DNA and base pairing to one of the two strands. This triple-stranded structure is called a displacement loop or D-loop.

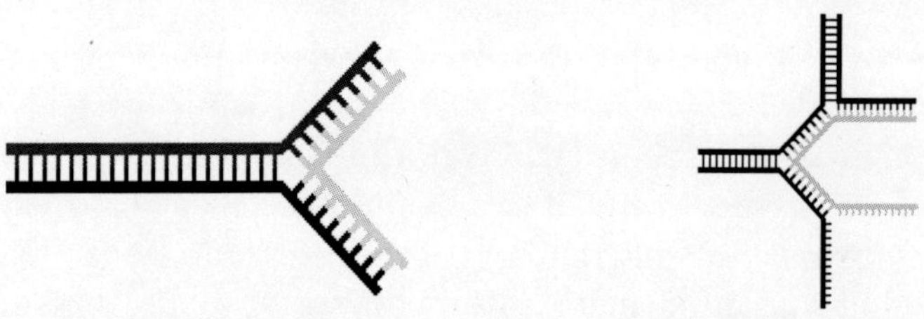

Single branch *Multiple branches*

Figure : *Branched DNA can form networks containing multiple branches*

Branched DNA

In DNA fraying occurs when non-complementary regions exist at the end of an otherwise complementary double-strand of DNA. However, branched DNA can occur if a third strand of DNA is introduced and contains adjoining regions able to hybridise with the frayed regions of the pre-existing double-strand. Although the simplest example of branched DNA involves only three strands of DNA, complexes involving additional strands and multiple branches are also possible. Branched DNA can be used in nanotechnology to construct geometric shapes.

Vibration

DNA may carry out low-frequency collective motion as observed by the Raman spectroscopy and analysed with a quasi-continuum model.

Chemical Modifications

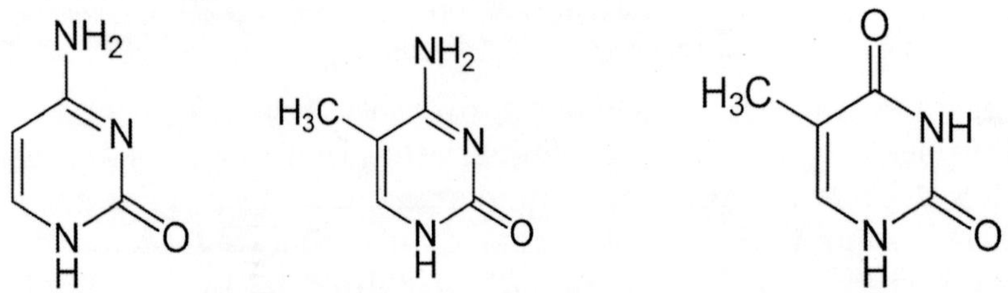

cytosine *5-methylcytosine* *thymine*

Structure of cytosine with and without the 5-methyl group. Deamination converts 5-methylcytosine into thymine.

Base Modifications

The expression of genes is influenced by how the DNA is packaged in chromosomes, in a structure called chromatin. Base modifications can be involved in packaging, with regions that have low or no gene expression usually containing high levels of methylation of cytosine bases. For example, cytosine methylation, produces 5-methylcytosine, which is important for X-chromosome inactivation.

The average level of methylation varies between organisms – the worm *Caenorhabditis elegans* lacks cytosine methylation, while vertebrates have higher levels, with up to 1% of their DNA containing 5-methylcytosine. Despite the importance of 5-methylcytosine, it can deaminate to leave a thymine base, so methylated cytosines are particularly prone to mutations. Other base modifications include adenine methylation in bacteria, the presence of 5-hydroxymethylcytosine in the brain, and the glycosylation of uracil to produce the “J-base” in kinetoplastids.

Damage

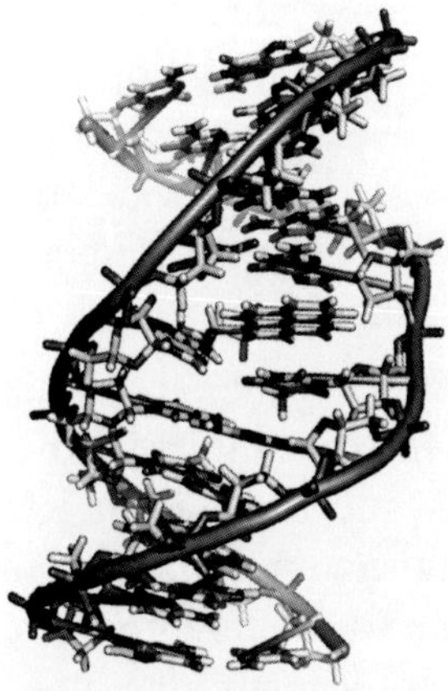

Figure : *A covalent adduct between a metabolically activated form of benzo[a]pyrene, the major mutagen in tobacco smoke, and DNA*

DNA can be damaged by many sorts of mutagens, which change the DNA sequence. Mutagens include oxidizing agents, alkylating agents and also high-energy electromagnetic radiation such as ultraviolet light and X-rays. The type of DNA damage produced depends on the type of mutagen.

For example, UV light can damage DNA by producing thymine dimers, which are cross-links between pyrimidine bases. On the other hand, oxidants such as free radicals or hydrogen peroxide produce multiple forms of damage, including base modifications, particularly of guanosine, and double-strand breaks.

A typical human cell contains about 150,000 bases that have suffered oxidative damage. Of these oxidative lesions, the most dangerous are double-strand breaks, as these are difficult to repair and can produce point mutations, insertions and deletions from the DNA sequence, as well as chromosomal translocations.

Many mutagens fit into the space between two adjacent base pairs, this is called *intercalation*. Most intercalators are aromatic and planar molecules; examples include ethidium bromide, acridines, daunomycin, and doxorubicin. In order for an intercalator to fit between base pairs, the bases must separate, distorting the DNA strands by unwinding of the double helix. This inhibits both transcription and DNA replication, causing toxicity and mutations.

As a result, DNA intercalators may be carcinogens, and in the case of thalidomide, a teratogen. Others such as benzo[*a*]pyrene diol epoxide and aflatoxin form DNA adducts which induce errors in replication. Nevertheless, due to their ability to inhibit DNA transcription and replication, other similar toxins are also used in chemotherapy to inhibit rapidly growing cancer cells.

Biological Functions

DNA usually occurs as linear chromosomes in eukaryotes, and circular chromosomes in prokaryotes. The set of chromosomes in a cell makes up its genome; the human genome has approximately 3 billion base pairs of DNA arranged into 46 chromosomes. The information carried by DNA is held in the sequence of pieces of DNA called genes. Transmission of genetic information in genes is achieved via complementary base pairing.

For example, in transcription, when a cell uses the information in a gene, the DNA sequence is copied into a complementary RNA sequence through the attraction between the DNA and the correct RNA nucleotides. Usually, this RNA copy is then used to make a matching protein sequence in a process called translation, which depends on the same interaction between RNA nucleotides. In alternative fashion, a cell may simply copy its genetic information in a process called DNA replication. The details of these functions are covered in other articles; here we focus on the interactions between DNA and other molecules that mediate the function of the genome.

Genes and Genomes

Genomic DNA is tightly and orderly packed in the process called DNA condensation to fit the small available volumes of the cell. In eukaryotes, DNA is located in the cell nucleus, as well as small amounts in mitochondria and chloroplasts. In prokaryotes, the DNA is held within an irregularly shaped body in the cytoplasm called the nucleoid. The genetic information in a genome is held within genes, and the complete set of this information in an organism is called its genotype. A gene is a unit of heredity and is a region of DNA that influences a particular characteristic in an organism. Genes contain an open reading frame that can be transcribed, as well as regulatory sequences such as promoters and enhancers, which control the transcription of the open reading frame.

In many species, only a small fraction of the total sequence of the genome encodes protein. For example, only about 1.5% of the human genome consists of protein-coding exons, with over 50% of human DNA consisting of non-coding repetitive sequences. The reasons for the presence of so much noncoding DNA in eukaryotic genomes and the extraordinary differences in genome size, or *C-value*, among species represent a long-standing puzzle known as the "C-value enigma". However, DNA sequences that do not code protein may still encode functional non-coding RNA molecules, which are involved in the regulation of gene expression.

Some noncoding DNA sequences play structural roles in chromosomes. Telomeres and centromeres typically contain few genes, but are important for the function and stability of chromosomes. An abundant form of noncoding

DNA in humans are pseudogenes, which are copies of genes that have been disabled by mutation. These sequences are usually just molecular fossils, although they can occasionally serve as raw genetic material for the creation of new genes through the process of gene duplication and divergence.

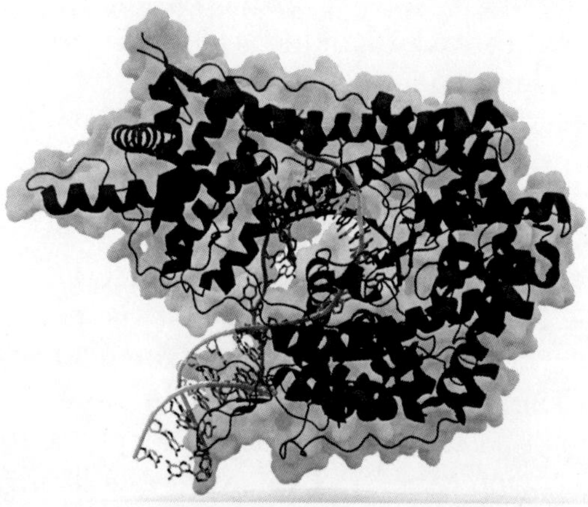

Figure : *T7 RNA polymerase (blue) producing a mRNA (green) from a DNA template (orange).*

Transcription and Translation

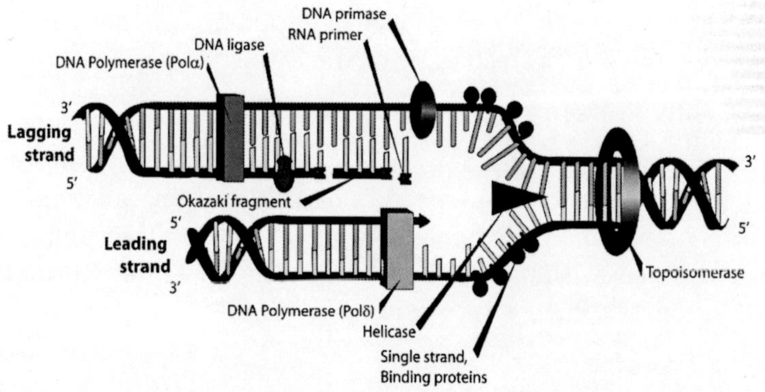

Figure : *DNA replication. The double helix is unwound by a helicase and topoisomerase. Next, one DNA polymerase produces the leading strand copy. Another DNA polymerase binds to the lagging strand. This enzyme makes discontinuous segments (called Okazaki fragments) before DNA ligase joins them together.*

A gene is a sequence of DNA that contains genetic information and can influence the phenotype of an organism. Within a gene, the sequence of bases along a DNA strand defines a messenger RNA sequence, which then defines one or more protein sequences. The relationship between the nucleotide sequences of genes and the amino-acid sequences of proteins is determined by the rules of translation, known collectively as the genetic code. The genetic code consists of three-letter 'words' called *codons* formed from a sequence of three nucleotides (e.g. ACT, CAG, TTT).In transcription, the codons of a gene are copied into messenger RNA by RNA polymerase. This RNA copy is then decoded by a ribosome that reads the RNA sequence by base-pairing the messenger RNA to transfer RNA, which carries amino acids. Since there are 4 bases in 3-letter combinations, there are 64 possible codons (4^3 combinations). These encode the twenty standard amino acids, giving most amino acids more than one possible codon. There are also three 'stop' or 'nonsense' codons signifying the end of the coding region; these are the TAA, TGA and TAG codons.

Replication

Cell division is essential for an organism to grow, but, when a cell divides, it must replicate the DNA in its genome so that the two daughter cells have the same genetic information as their parent. The double-stranded structure of DNA provides a simple mechanism for DNA replication. Here, the two strands are separated and then each strand's complementary DNA sequence is recreated by an enzyme called DNA polymerase. This enzyme makes the complementary strand by finding the correct base through complementary base pairing, and bonding it onto the original strand. As DNA polymerases can only extend a DNA strand in a 52 to 32 direction, different mechanisms are used to copy the antiparallel strands of the double helix. In this way, the base on the old strand dictates which base appears on the new strand, and the cell ends up with a perfect copy of its DNA.

Interactions with Proteins

All the functions of DNA depend on interactions with proteins. These protein interactions can be non-specific, or the protein can bind specifically to a single DNA sequence. Enzymes can also bind to DNA and of these, the polymerases that copy the DNA base sequence in transcription and DNA replication are particularly important.

DNA-binding Proteins

Structural proteins that bind DNA are well-understood examples of non-specific DNA-protein interactions. Within chromosomes, DNA is held in complexes with structural proteins. These proteins organise the DNA into a compact structure called chromatin. In eukaryotes this structure involves DNA binding to a complex of small basic proteins called histones, while in prokaryotes multiple types of proteins are involved. The histones form a disk-shaped complex called a nucleosome, which contains two complete turns of double-stranded DNA wrapped around its surface. These non-specific interactions are formed through basic residues in the histones making ionic bonds to the acidic sugar-phosphate backbone of the DNA, and are therefore largely independent of the base sequence.

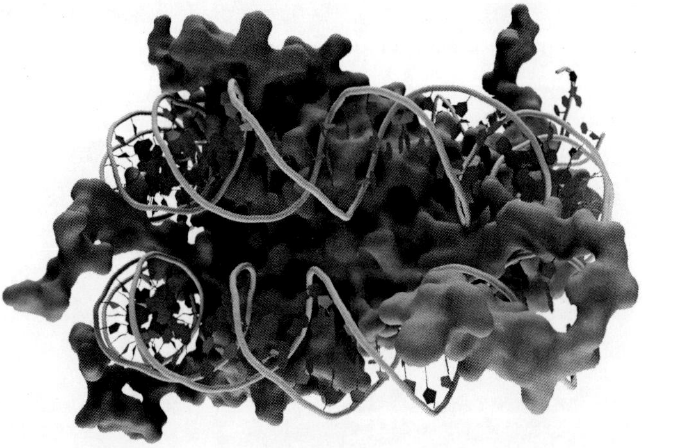

Figure: *Interaction of DNA (shown in orange) with histones (shown in blue). These proteins' basic amino acids bind to the acidic phosphate groups on DNA.*

Chemical modifications of these basic amino acid residues include methylation, phosphorylation and acetylation. These chemical changes alter the strength of the interaction between the DNA and the histones, making the DNA more or less accessible to transcription factors and changing the rate of transcription. Other non-specific DNA-binding proteins in chromatin include the high-mobility group proteins, which bind to bent or distorted DNA.

These proteins are important in bending arrays of nucleosomes and arranging them into the larger structures that make up chromosomes. A distinct group of DNA-binding proteins are the DNA-binding proteins that specifically bind single-stranded DNA. In humans, replication protein A is the best-understood member of this family and is used in processes where the double helix is separated, including DNA replication, recombination and DNA repair. These binding proteins seem to stabilize single-stranded DNA and protect it from forming stem-loops or being degraded by nucleases. In contrast, other proteins have evolved to bind to particular DNA sequences. The most intensively studied of these are the various transcription factors, which are proteins that regulate transcription. Each transcription factor binds to one particular set of DNA sequences and activates or inhibits the transcription of genes that have these sequences close to their promoters.

The transcription factors do this in two ways. Firstly, they can bind the RNA polymerase responsible for transcription, either directly or through other mediator proteins; this locates the polymerase at the promoter and allows it to begin transcription. Alternatively, transcription factors can bind enzymes that modify the histones at the promoter; this will change the accessibility of the DNA template to the polymerase. As these DNA targets can occur throughout an organism's genome, changes in the activity of one type of transcription factor can affect thousands of genes. Consequently, these proteins are often the targets of the signal transduction processes that control responses to environmental changes or cellular differentiation and development. The specificity of these transcription factors' interactions with DNA come from the proteins making multiple contacts to the edges of the DNA bases, allowing them to "read" the DNA sequence. Most of these base-interactions are made in the major groove, where the bases are most accessible.

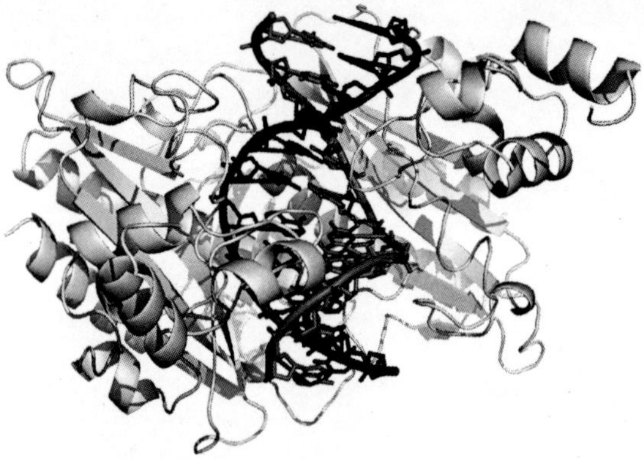

Figure : *The restriction enzyme EcoRV (green) in a complex with its substrate DNA*

DNA-modifying Enzymes

Nucleases and Ligases

Nucleases are enzymes that cut DNA strands by catalysing the hydrolysis of the phosphodiester bonds. Nucleases that hydrolyse nucleotides from the ends of DNA strands are called exonucleases, while endonucleases cut within strands. The most frequently used nucleases in molecular biology are the restriction endonucleases, which cut DNA at specific sequences. For instance, the EcoRV enzyme shown to the left recognises the 6-base sequence 52 - GAT | ATC-32 and makes a cut at the vertical line. In nature, these enzymes protect bacteria against phage infection by digesting the phage DNA when it enters the bacterial cell, acting as part of the restriction modification system. In technology, these sequence-specific nucleases are used in molecular cloning and DNA fingerprinting. Enzymes called DNA ligases can rejoin cut or broken DNA strands. Ligases are particularly important in lagging strand DNA replication, as they join together the short segments of DNA produced at the replication fork into a complete copy of the DNA template. They are also used in DNA repair and genetic recombination.

Topoisomerases and Helicases

Topoisomerases are enzymes with both nuclease and ligase activity. These proteins change the amount of supercoiling in DNA. Some of these enzymes work by cutting the DNA helix and allowing one section to rotate, thereby reducing its level of supercoiling; the enzyme then seals the DNA break. Other types of these enzymes are capable of cutting one DNA helix and then passing a second strand of DNA through this break, before rejoining the helix. Topoisomerases are required for many processes involving DNA, such as DNA replication and transcription. Helicases are proteins that are a type of molecular motor. They use the chemical energy in nucleoside triphosphates, predominantly ATP, to break hydrogen bonds between bases and unwind the DNA double helix into single strands. These enzymes are essential for most processes where enzymes need to access the DNA bases.

Polymerases

Polymerases are enzymes that synthesize polynucleotide chains from nucleoside triphosphates. The sequence of their products are copies of existing polynucleotide chains – which are called *templates*. These enzymes function by adding nucleotides onto the 32 hydroxyl group of the previous nucleotide in a DNA strand. As a consequence, all polymerases work in a 52 to 32 direction. In the active site of these enzymes, the incoming nucleoside triphosphate base-pairs to the template: this allows polymerases to accurately synthesize the complementary strand of their template. Polymerases are classified according to the type of template that they use.

In DNA replication, a DNA-dependent DNA polymerase makes a copy of a DNA sequence. Accuracy is vital in this process, so many of these polymerases have a proofreading activity. Here, the polymerase recognises the occasional mistakes in the synthesis reaction by the lack of base pairing between the mismatched nucleotides. If a mismatch is detected, a 32 to 52 exonuclease activity is activated and the incorrect base removed. In most organisms, DNA polymerases function in a large complex called the replisome that contains multiple accessory subunits, such as the DNA clamp or helicases. RNA-dependent DNA polymerases are a specialised class of polymerases that copy the sequence of an RNA strand into DNA.

They include reverse transcriptase, which is a viral enzyme involved in the infection of cells by retroviruses, and telomerase, which is required for the replication of telomeres. Telomerase is an unusual polymerase because it contains its own RNA template as part of its structure. Transcription is carried out by a DNA-dependent RNA polymerase that copies the sequence of a DNA strand into RNA. To begin transcribing a gene, the RNA polymerase binds to a sequence of DNA called a promoter and separates the DNA strands. It then copies the gene sequence into a messenger RNA transcript until it reaches a region of DNA called the terminator, where it halts and detaches from the DNA. As with human DNA-dependent DNA polymerases, RNA polymerase II, the enzyme that transcribes most of the genes in the human genome, operates as part of a large protein complex with multiple regulatory and accessory subunits.

Genetic Recombination

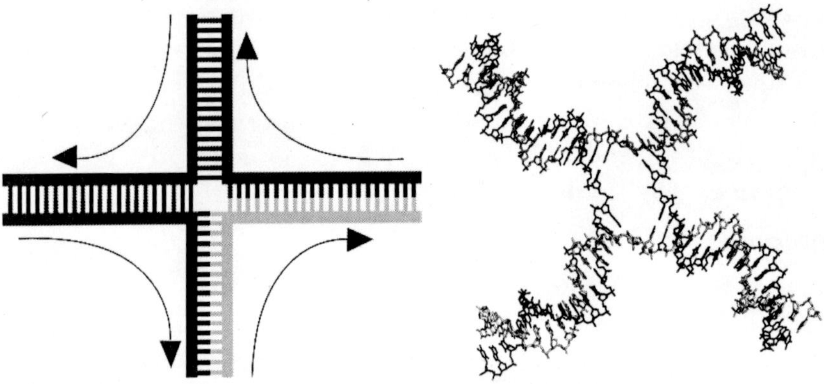

Figure : Structure of the Holliday junction intermediate in genetic recombination. The four separate DNA strands are coloured red, blue, green and yellow.

A DNA helix usually does not interact with other segments of DNA, and in human cells the different chromosomes even occupy separate areas in the nucleus called "chromosome territories". This physical separation of different chromosomes is important for the ability of DNA to function as a stable repository for information, as one of the few times chromosomes interact is during chromosomal crossover when they recombine. Chromosomal crossover is when two DNA helices break, swap a section and then rejoin. Recombination allows chromosomes to exchange genetic information and produces new combinations of genes, which increases the efficiency of natural selection and can be important in the rapid evolution of new proteins. Genetic recombination can also be involved in DNA repair, particularly in the cell's response to double-strand breaks. The most common form of chromosomal crossover is homologous recombination, where the two chromosomes involved share very similar sequences. Non-homologous recombination can be damaging to cells, as it can produce chromosomal translocations and genetic abnormalities. The recombination reaction is catalysed by enzymes known as recombinases, such as RAD51. The first step in recombination is a double-stranded break either caused by an endonuclease or damage to the DNA. A series of steps catalysed in part by the recombinase then leads to joining of the two helices by at least one Holliday junction, in which a segment of a single strand in each helix is annealed to the complementary strand in the other helix. The Holliday junction is a tetrahedral junction structure that can be moved along the pair of chromosomes, swapping one strand for another. The recombination reaction is then halted by cleavage of the junction and re-ligation of the released DNA.

Evolution

DNA contains the genetic information that allows all modern living things to function, grow and reproduce. However, it is unclear how long in the 4-billion-year history of life DNA has performed this function, as it has been proposed that the earliest forms of life may have used RNA as their genetic material.

RNA may have acted as the central part of early cell metabolism as it can both transmit genetic information and carry out catalysis as part of ribozymes. This ancient RNA world where nucleic acid would have been used for both catalysis and genetics may have influenced the evolution of the current genetic code based on four nucleotide bases. This would occur, since the number of different bases in such an organism is a trade-off between a small number of bases increasing replication accuracy and a large number of bases increasing the catalytic efficiency of ribozymes.

However, there is no direct evidence of ancient genetic systems, as recovery of DNA from most fossils is impossible. This is because DNA will survive in the environment for less than one million years and slowly degrades into short fragments in solution. Claims for older DNA have been made, most notably a report of the isolation of a viable bacterium from a salt crystal 250 million years old, but these claims are controversial. On August 8, 2011, a report, based on NASA studies with meteorites found on Earth, was published suggesting building blocks of DNA (adenine, guanine and related organic molecules) may have been formed extraterrestrially in outer space.

Uses in Technology

Genetic Engineering

Methods have been developed to purify DNA from organisms, such as phenol-chloroform extraction, and to manipulate it in the laboratory, such as restriction digests and the polymerase chain reaction. Modern biology and biochemistry make intensive use of these techniques in recombinant DNA technology. Recombinant DNA is a man-made DNA sequence that has been assembled from other DNA sequences. They can be transformed into organisms in the form of plasmids or in the appropriate format, by using a viral vector. The genetically modified organisms produced can be used to produce products such as recombinant proteins, used in medical research, or be grown in agriculture.

Forensics

Forensic scientists can use DNA in blood, semen, skin, saliva or hair found at a crime scene to identify a matching DNA of an individual, such as a perpetrator. This process is formally termed DNA profiling, but may also be called "genetic fingerprinting". In DNA profiling, the lengths of variable sections of repetitive DNA, such as short tandem repeats and minisatellites, are compared between people. This method is usually an extremely reliable technique for identifying a matching DNA. However, identification can be complicated if the scene is contaminated with DNA from several people. DNA profiling was developed in 1984 by British geneticist Sir Alec Jeffreys, and first used in forensic science to convict Colin Pitchfork in the 1988 Enderby murders case.

People convicted of certain types of crimes may be required to provide a sample of DNA for a database. This has helped investigators solve old cases where only a DNA sample was obtained from the scene. DNA profiling can also be used to identify victims of mass casualty incidents. On the other hand, many convicted people have been released from prison on the basis of DNA techniques, which were not available when a crime had originally been committed.

Bioinformatics

Bioinformatics involves the manipulation, searching, and data mining of biological data, and this includes DNA sequence data. The development of techniques to store and search DNA sequences have led to widely applied advances in computer science, especially string searching algorithms, machine learning and database theory. String searching or matching algorithms, which find an occurrence of a sequence of letters inside a larger sequence of letters, were developed to search for specific sequences of nucleotides. The DNA sequenced may be aligned with other DNA sequences to identify homologous sequences and locate the specific mutations that make them distinct. These techniques, especially multiple sequence alignment, are used in studying phylogenetic relationships and protein function. Data sets representing entire genomes' worth of DNA sequences, such as those produced by the Human Genome Project, are difficult to use without the annotations that identify the locations of genes and regulatory elements on each chromosome.

Regions of DNA sequence that have the characteristic patterns associated with protein- or RNA-coding genes can be identified by gene finding algorithms, which allow researchers to predict the presence of particular gene products and their possible functions in an organism even before they have been isolated experimentally. Entire genomes may also be compared which can shed light on the evolutionary history of particular organism and permit the examination of complex evolutionary events.

DNA Nanotechnology

DNA nanotechnology uses the unique molecular recognition properties of DNA and other nucleic acids to create self-assembling branched DNA complexes with useful properties. DNA is thus used as a structural material rather than as a carrier of biological information. This has led to the creation of two-dimensional periodic lattices (both tile-based as well as using the "DNA origami" method) as well as three-dimensional structures in the shapes of polyhedra. Nanomechanical devices and algorithmic self-assembly have also been demonstrated, and these DNA structures have been used to template the arrangement of other molecules such as gold nanoparticles and streptavidin proteins.

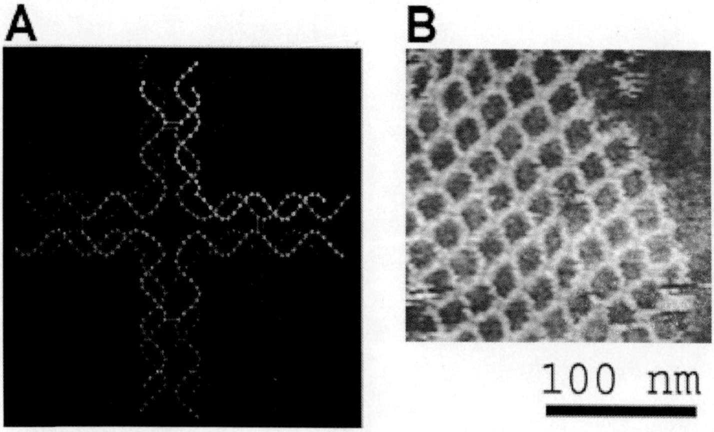

Figure : *The DNA structure at left (schematic shown) will self-assemble into the structure visualised by atomic force microscopy at right. DNA nanotechnology is the field that seeks to design nanoscale structures using the molecular recognition properties of DNA molecules. Image from Strong, 2004.*

Because DNA collects mutations over time, which are then inherited, it contains historical information, and, by comparing DNA sequences, geneticists can infer the evolutionary history of organisms, their phylogeny. This field of phylogenetics is a powerful tool in evolutionary biology. If DNA sequences within a species are compared, population geneticists can learn the history of particular populations. This can be used in studies ranging from ecological genetics to anthropology; For example, DNA evidence is being used to try to identify the Ten Lost Tribes of Israel. DNA has also been used to look at modern family relationships, such as establishing family relationships between the descendants of Sally Hemings and Thomas Jefferson. This usage is closely related to the use of DNA in criminal investigations detailed above. Indeed, some criminal investigations have been solved when DNA from crime scenes has matched relatives of the guilty individual.

History of DNA Research

DNA was first isolated by the Swiss physician Friedrich Miescher who, in 1869, discovered a microscopic substance in the pus of discarded surgical bandages. As it resided in the nuclei of cells, he called it "nuclein". In 1878,

Albrecht Kossel isolated the non-protein component of "nuclein", nucleic acid, and later isolated its five primary nucleobases. In 1919, Phoebus Levene identified the base, sugar and phosphate nucleotide unit. Levene suggested that DNA consisted of a string of nucleotide units linked together through the phosphate groups.

However, Levene thought the chain was short and the bases repeated in a fixed order. In 1937 William Astbury produced the first X-ray diffraction patterns that showed that DNA had a regular structure.

In 1927 Nikolai Koltsov proposed that inherited traits would be inherited via a "giant hereditary molecule" which would be made up of "two mirror strands that would replicate in a semi-conservative fashion using each strand as a template". In 1928, Frederick Griffith discovered that traits of the "smooth" form of the *Pneumococcus* could be transferred to the "rough" form of the same bacteria by mixing killed "smooth" bacteria with the live "rough" form. T

his system provided the first clear suggestion that DNA carries genetic information—the Avery–MacLeod–McCarty experiment—when Oswald Avery, along with coworkers Colin MacLeod and Maclyn McCarty, identified DNA as the transforming principle in 1943. DNA's role in heredity was confirmed in 1952, when Alfred Hershey and Martha Chase in the Hershey–Chase experiment showed that DNA is the genetic material of the T2 phage.

In 1953, James D. Watson and Francis Crick suggested what is now accepted as the first correct double-helix model of DNA structure in the journal *Nature*. Their double-helix, molecular model of DNA was then based on a single X-ray diffraction image taken by Rosalind Franklin and Raymond Gosling in May 1952, as well as the information that the DNA bases are paired — also obtained through private communications from Erwin Chargaff in the previous years.

Chargaff's rules played a very important role in establishing double-helix configurations for B-DNA as well as A-DNA. Experimental evidence supporting the Watson and Crick model were published in a series of five articles in the same issue of *Nature*. Of these, Franklin and Gosling's paper was the first publication of their own X-ray diffraction data and original analysis method that partially supported the Watson and Crick model; this issue also contained an article on DNA structure by Maurice Wilkins and two of his colleagues, whose analysis and *in vivo* B-DNA X-ray patterns also supported the presence *in vivo* of the double-helical DNA configurations as proposed by Crick and Watson for their double-helix molecular model of DNA in the previous two pages of *Nature*.

In 1962, after Franklin's death, Watson, Crick, and Wilkins jointly received the Nobel Prize in Physiology or Medicine. However, Nobel rules of the time allowed only living recipients, but a vigorous debate continues on who should receive credit for the discovery.

In an influential presentation in 1957, Crick laid out the central dogma of molecular biology, which foretold the relationship between DNA, RNA, and proteins, and articulated the "adaptor hypothesis".

Final confirmation of the replication mechanism that was implied by the double-helical structure followed in 1958 through the Meselson–Stahl experiment.

Further work by Crick and coworkers showed that the genetic code was based on non-overlapping triplets of bases, called codons, allowing Har Gobind Khorana, Robert W. Holley and Marshall Warren Nirenberg to decipher the genetic code.

These findings represent the birth of molecular biology.

4

Genetic Mechanisms

In addition to very frequent simple mutations, there are several mutation classes which involve sequence exchange between allelic or nonallelic sequences, often involving repeated sequences. For example, tandemly repetitive DNA is prone to deletion/insertion polymorphism whereby different alleles vary in the number of integral copies of the tandem repeat. Such variable number of tandem repeat (VNTR) polymorphisms can occur in the case of repeated units that are very short (microsatellites); intermediate (minisatellites) or large. Different genetic mechanisms can account for VNTR polymorphism depending on the size of the repeating unit. In addition, interspersed repeats can also predispose to deletions/duplications by a variety of different genetic mechanisms.

Slipped Strand Mispairing can Cause VNTR Polymorphism at Short Tandem Repeats (Microsatellites)

There is considerable variation in the germline mutation rates at microsatellite loci, ranging from an undetectable level up to about 8×10^{-3}. Novel length alleles at (CA)/ (TG) microsatellites and at tetranucleotide marker loci are known to be formed without exchange of flanking markers. This means that they are not generated by unequal crossover. Instead, as new mutant alleles have been observed to differ by a single repeat unit from the originating parental allele (Mahtani and Willard, 1993), the most likely mechanism to explain length variation is a form of exchange of sequence information which commences by slipped strand mispairing.

This occurs when the normal pairing between the two complementary strands of a double helix is altered by staggering of the repeats on the two strands, leading to incorrect pairing of repeats. Although slipped strand mispairing can be envisaged to occur in nonreplicating DNA, replicating DNA may offer more opportunity for slippage and hence the mechanism is often also called replication slippage or polymerase slippage. In addition to mispairing between tandem repeats, slippage replication has been envisaged to generate large deletions and duplications by mispairing between noncontiguous repeats and has been suggested to be a major mechanism for DNA sequence and genome evolution. The pathogenic potential of short tandem repeats is considerable.

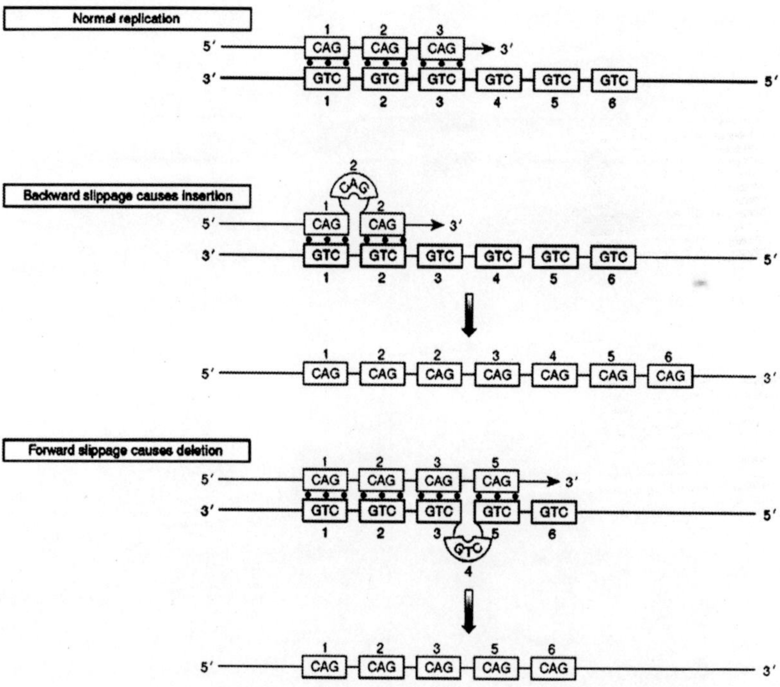

Figure: *Slipped strand mispairing during DNA replication can cause insertions or deletions.*

Large Units of Tandemly Repeated DNA are Prone to Insertion/deletion as a Result of Unequal Crossover or Unequal Sister Chromatid Exchanges

Homologous recombination describes recombination (crossover) occurring at meiosis or, rarely, mitosis between identical or very similar DNA sequences. It usually involves breakage of *nonsister chromatids* of a pair of homologs and rejoining of the fragments to generate new recombinant strands. Sister chromatid exchange is an analogous type of sequence exchange involving breakage of individual *sister chromatids* and rejoining fragments that initially were on different chromatids of the same chromosome. Both homologous recombination and sister chromatid exchange normally involve *equal* exchanges

- cleavage and rejoining of the chromatids occurs at the same position on each chromatid. As a result, the exchanges occur *between allelic sequences* and at corresponding positions within alleles. In the case of intragenic equal crossover between two alleles, a new allele can result which is a fusion gene (or hybrid gene), comprising a terminal fragment from one allele and the remaining sequence of the second allele. However, equal sister chromatid exchanges cannot normally produce genetic variation because sister chromatids have identical DNA sequences.

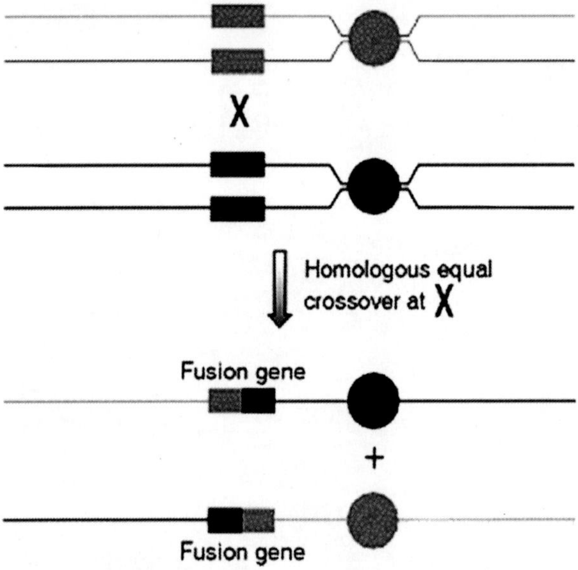

Figure: *Homologous equal crossover can result in fusion genes.*

Unequal crossover is a form of recombination in which the crossover takes place *between nonallelic sequences* on nonsister chromatids of a pair of homologs. Often the sequences at which crossover takes place show very considerable sequence homology which presumably stabilizes mispairing of the chromosomes. Because crossover occurs between mispaired nonsister chromatids, the exchange results in a deletion on one of the participating chromatids and an insertion on the other. The analogous exchange between sister chromatids is called unequal sister chromatid exchange. Both mechanisms occur predominantly at locations where the tandemly repeated units are moderate to large in size. In such cases, the very high degree of sequence homology between the different repeats can facilitate pairing of nonallelic repeats on nonsister chromatids or sister chromatids. If chromosome breakage and rejoining occurs while the chromatids are mispaired in this way, an insertion or deletion of an integral number of repeat units will result. Note that such exchanges are *reciprocal*; both participating chromatids are modified, in one case resulting in an insertion, and in the other case in a complementary deletion.

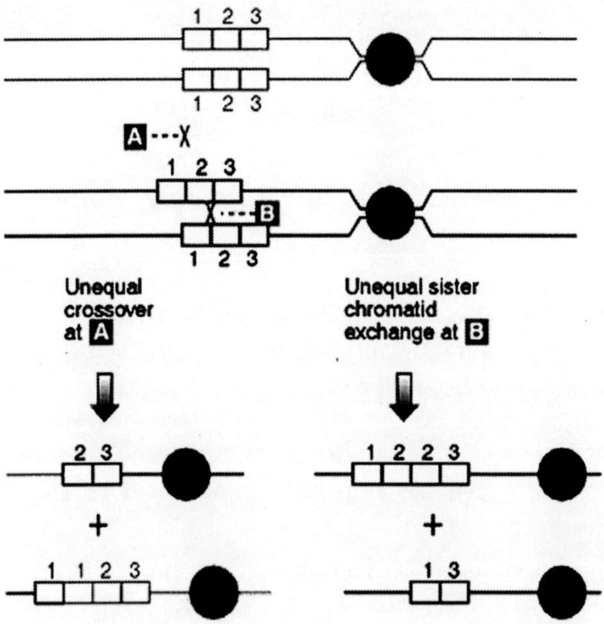

Figure: *Unequal crossover and unequal sister chromatid exchange cause insertions and deletions.*

Unequal sister chromatid exchange is thought to be a major mechanism underlying VNTR polymorphism in the rDNA clusters. Unequal crossover is also expected to occur comparatively frequently in complex satellite DNA repeats and at tandemly repeated gene loci. In the latter case, unequal crossover is known to generate pathogenic deletions at some loci.

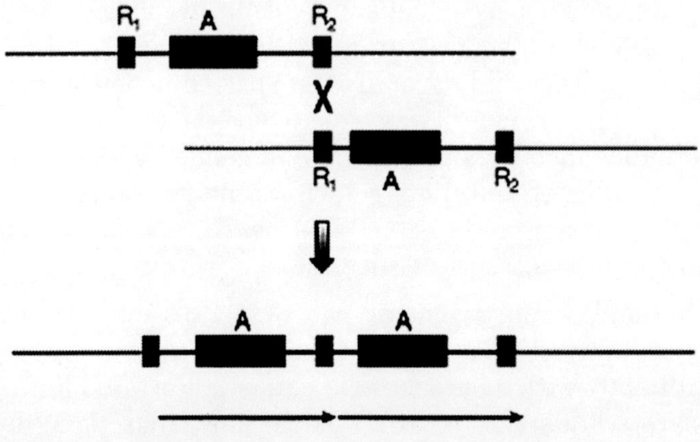

Figure: *Tandem gene duplication can result from unequal crossover or unequal sister chromatid exchange, facilitated by short interspersed repeats.*

Such exchanges can also lead to concerted evolution by causing a particular variant to spread through an array of tandem repeats, resulting in *homogenisation* of the repeat units. Occasionally, unequal crossover and unequal sister chromatid exchanges can occur at regions where there is little homology. This is likely to be the case when such mechanisms first generate a tandemly duplicated locus following mispairing of nonallelic repeats such as two *Alu* repeats or even smaller elements.

Gene Conversion Events may be Relatively Frequent in Tandemly Repetitive DNA

Gene conversion describes a *nonreciprocal* transfer of sequence information between a pair of nonallelic DNA sequences (*interlocus gene conversion)* or allelic sequences (*interallelic gene conversion*). One of the pair of interacting sequences, the *donor*, remains unchanged. The other DNA sequence, the *acceptor*, is changed by having some or all of its sequence replaced by a sequence copied from the donor sequence. The sequence exchange is therefore a directional one; the acceptor sequence is modified by the donor sequence, but not the other way round.

One possible mechanism for gene conversion envisages formation of a heteroduplex between a DNA strand from the donor gene and a complementary strand from the acceptor gene. Following heteroduplex formation, conversion of an acceptor gene segment may occur by mismatch repair - DNA repair enzymes recognize that the two strands of the heteroduplex are not perfectly matched and 'correct' the DNA sequence of the acceptor strand to make it perfectly complementary in the converted region to the sequence of the donor gene strand. Gene conversion has been well-described in fungi where all four products of meiosis can be recovered and studied (tetrad analysis). In humans and mammals it is not possible to do this and so gene conversion cannot be demonstrated unambiguously in higher organisms (it can never be distinguished from double crossover events, for example, although double crossovers occurring in very close proximity would normally be expected to be extremely unlikely). Despite the difficulty in identifying gene conversion in complex organisms, there are numerous instances in mammalian genomes where an allele at one locus shows a pattern of mutations which strongly resembles those found in alleles at another locus of the same species. Such evidence suggests gene conversion-like exchanges between loci.

Although simple comparisons of two sequences may be suggestive, the evidence for gene conversion is most compelling when a new mutant allele can be compared directly with its progenitor sequence. Certain highly mutable loci lend themselves to this type of analysis. In particular, some hypervariable minisatellite loci have high germline mutation rates (often 1% or more per gamete) and individual repeats often show nucleotide differences so that repeat subclasses can be recognised. Germline mutations can be studied by detecting and characterising mutant minisatellite alleles in individual gametes.

To do this, PCR analysis has been conducted on multiple dilute aliquots of DNA isolated from the sperm of an individual (small pool PCR), where each aliquot is calibrated to contain a few, perhaps 100, input molecules (Jeffreys *et al.,* 1994).

The PCR products recovered from individual pools can then be typed to identify any new mutations that result in a novel allele whose length is sufficiently different as to be distinguishable from the progenitor allele. Analyses of the patterns of germline mutation at three such loci have failed to identify exchanges of flanking markers and have shown that most mutations occurring at these loci are polar, involving the preferential gain of a few repeats at one end of a tandem repeat array. There is a bias towards gain of repeats and evidence was obtained for nonreciprocal sequence exchange between alleles, suggesting interallelic gene conversion (Jeffreys *et al.,* 1994). Evidence for interlocus gene conversion has also been obtained in human genes, notably the steroid 21-hydroxylase gene.

Pathogenic Mutations

There is a High Deleterious Mutation Rate in Hominids

Neutral mutations (those which are neither detrimental nor advantageous for the organism carrying them) accumulate throughout the generations at a rate equal to the mutation rate. To get an estimate of the (total) mutation rate is therefore simple: all one needs to do is to measure the rate of change of some presumed neutral sequence (e.g. intronic, pseudogene, etc). The deleterious mutation rate, by contrast, has been notoriously difficult to measure and no convincing estimate existed for any vertebrate until a study reported by Eyre-Walker and Keightley (1999). They investigated amino acid changes in 46 proteins occurring in the human ancestral line after its divergence from the chimpanzee. If all non-synonymous substitutions were neutral, 231 new substitutions would have been expected in their sample of genes (given an average neutral mutation rate of 0.0056 nonsynonymous substitutions per nucleotide and a total of 41 471 nucleotides investigated).

Instead, only 143 nonsynonymous substitutions were observed and 88 such substitutions were inferred to have been removed by natural selection because they had been deleterious. On the assumption of 60 000 genes, and 240 000 generations since human-chimpanzee divergence, they estimated a deleterious rate of 1.6 mutations per person per generation out of a total of 4.2 mutations per person generation. The estimated deleterious mutation rates in chimpanzees and gorillas were very similar but those for rodent-specific lineages are about one order of magnitude less, possibly because of much smaller numbers of germ cell divisions in rodents. The very high deleterious rate in hominids may even be an underestimate. If the total gene number were

80 000 and an average coding sequence was 1800 nucleotides, the estimated deleterious mutation rate would be 2.5 mutations per person per generation and, on other grounds too, a more likely rate has been considered to be three deleterious mutations per person per generation (Crow, 1999).

So with three genetic deaths per person why are we not extinct? The data would suggest that harmful mutations need to be weeded out in clusters at a time. One way to achieve this would be if natural selection operated such that individuals with the most mutations are preferentially eliminated (e.g. harmful mutations interact). This could only happen in a sexual species where mutations are shuffled each generation by genetic recombination, and so the existence of such a high deleterious mutation rate has been taken as further vindication that sex (meiotic recombination) is an efficient way to eliminate harmful mutations.

Pathogenic Mutations are Preferentially Located at Certain Types of Intragenic DNA Sequence

Pathogenic mutations can occur at three types of DNA sequence at a gene locus.

- *The coding sequence of the gene.* This is where the great majority of recorded pathogenic mutations have been identified. Those due to nucleotide substitution are, in the vast majority of cases, nonsynonymous substitutions and mostly occur at first and second base positions of codons. However, very rarely, a synonymous codon substitution is not neutral as expected, but may cause disease by activating a cryptic splice site. Because of its relatively high mutability, the CpG dinucleotide is often located at hotspots for pathogenic mutation in coding DNA (Cooper and Youssoufian, 1988). Other hotspots include tandem repeats within coding DNA.
- *Intragenic noncoding sequences.* This is restricted to sequences which are necessary for correct expression of the gene, such as important intronic elements, notably the highly conserved GT and AG dinucleotides at the ends of introns, but also conserved elements of the untranslated sequences. Often such mutations represent a small component (~10–15%) of the total pathogenic mutations at a gene locus (Cooper *et al.*, 1995). However, in some disorders pathogenic splicing mutations may be common. In the case of the collagen disorder osteogenesis imperfecta they constitute a very common pathological mutation which is second in frequency only to substitutions leading to replacement of the highly conserved, structurally important glycine residues. The collagen genes have small exons and a comparatively large number of introns (often more than 50 and as many as 106 in the case of the *COL7A1* gene), making them exceptional targets for splicing mutations. Occasional

pathogenic mutations have been recorded in the 52 UTR (such as in the case of hemophilia B Leyden) and appear to exert their effect at the transcriptional level. Several examples are also known of pathogenic mutations in the 32 UTR.

- *Regulatory sequences outside exons.* Most mutations located in regulatory sequences have been identified in conserved elements located just upstream of the first exon, notably promoter elements. In addition, other more distantly located regulatory elements may be sites of pathological mutation. For example, deletions which eliminate the â-globin LCR but leave the â-globin gene and its promoter intact result in almost complete abolition of â-globin gene expression and contribute to â-thalassemia. Clearly, in some cases a gene may be regulated by the product of a distantly related gene. For example, in the case of rare variants of α-thalassemia with mental retardation, the α-globin gene and its promoter may show no evidence of pathological mutation and the disease maps to an X-linked gene which encodes a transcription factor, one of whose target sequences is presumably the α-globin gene (Gibbons *et al.*, 1995).

The Mitochondrial Genome is a Hotspot for Pathogenic Mutations

Because of the very large size of the human nuclear genome, most mutations occur in nuclear DNA sequences. By comparison, the mitochondrial genome is a small target for mutation (about 1/200 000 of the size of the nuclear genome). Unlike nuclear genes, mitochondrial genes are present in numerous copies (there are thousands of copies of the mtDNA molecule in each human somatic cell; some cells, such as brain and muscle cells, have particularly high oxidative phosphorylation requirements and so more mitochondria). The mtDNA is inherited from the maternal oocyte, which is an exceptional cell with many more mtDNA molecules than somatic cells. Given that a mutation in mitochondrial DNA must arise on a single mtDNA molecule, one might intuitively expect that the chances of a single mtDNA mutation becoming *fixed* would be very low and the mutation rate correspondingly low.

On these grounds, one could anticipate that the proportion of clinical disease due to pathogenic mutation in the mitochondrial genome should be extremely low. Instead, the frequency of 'mitochondrial disorders' is rather high and the mitochondrial genome can be considered to be a mutation hotspot. Different factors can explain this apparent paradox:

- *Differential target size for pathogenic mutation.* Pathogenesis is associated with mutations in coding DNA and the mitochondrial genome has a much higher percentage of coding DNA (93%) than found in the nuclear genome (3%). When this is taken into consideration, however, there is still a large imbalance: about 100 Mb of coding DNA in the nuclear genome but only 15.4 kb of coding sequence in the mitochondrial genome, giving a target ratio of 6000:1 in favour of the nuclear genome.

- *High mutation rate in mtDNA.* The mitochondrial genome is much more prone to nucleotide change than the nuclear genome. Even although about 100 000 copies of the mitochondrial genome are maternally inherited in the fertilized oocyte there are mechanisms which permit rapid fixation of mutations in mitochondrial DNA. The combination of mtDNA instability and a high fixation rate means that the mutation rate in mitochondrial DNA is very high. Mutations have been reported to be fixed in the mitochondrial genomes of animal cells at a rate which is about 10 times greater than occurring in equivalent sequences in the nuclear genome (Brown *et al.*, 1979). This means that the small recombination-deficient animal mtDNA molecules appear to be evolving remarkably rapidly, corresponding to about 2–4% sequence divergence per million years. In contrast, plant mtDNA molecules are comparatively large (150 kb-2.5 Mb), have introns, engage in recombination and are evolving comparatively slowly.

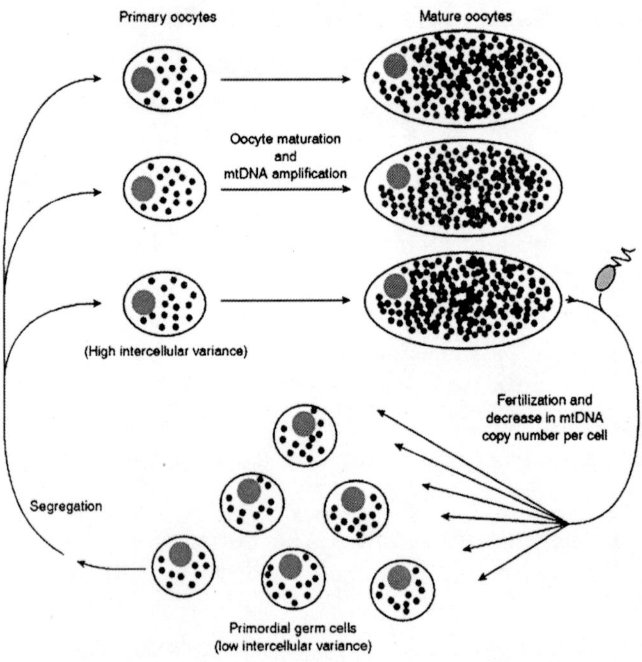

Figure: *Segregation of mitochondrial genotypes in the maternal germ line.*

The high instability of mtDNA has been postulated to result from several factors. The high rate of production of reactive oxygen intermediates by the respiratory chain is thought to cause substantial oxidative damage to mtDNA which, unlike nuclear DNA, is not protected by histones. The mtDNA also has to undergo many more rounds of replication than chromosomal DNA. Although several well-characterised mtDNA repair systems are now known, some frequent mutations cannot be repaired, including thymidine dimers.

Expression of Pathogenic Mutations

The degree to which a pathogenic mutation results in an aberrant phenotype depends on several factors:

- *The mutation class and the way in which the expression of the mutant gene is altered.* This may depend on the location of the mutation within the gene. Many pathogenic mutations result in abolition or substantial reduction of gene expression, but some lead to inappropriate expression. For example, overexpression of a gene product may cause an abnormal phenotype where gene dosage needs to be carefully regulated, and ectopic expression, that is expression in tissues where the gene is not normally expressed, may also be harmful.
- *The degree to which aspects of the aberrant phenotype are expressed in the heterozygote.* The presence of a single normal allele may be sufficient to maintain a clinically normal phenotype (as in recessively inherited disorders), or a milder phenotype when compared with that of mutant homozygotes, as in dominantly inherited disorders where the mutation is a simple *loss of function* mutation.
- *The degree to which expression of a mutant phenotype is influenced by other gene products.* The same mutant allele can have different phenotypic effects on different genetic backgrounds, depending on particular alleles at other gene loci (*modifier genes*).
- *The proportion and nature of cells in which the mutant gene is present.* Generally, mutations which are present in all the cells of an individual (inherited mutations) or in many of them (somatic mutations acquired very early in development) are likely to have a more profound effect than those present in a few cells (somatic mutations which arise at much later stages) or in cell types where the relevant gene is not expressed. Cancers, however, arise from unregulated division of cells produced from a single original mutant cell.
- *The parental origin of the mutation.* This is only known to be important in the case of the few genes which are *imprinted*.

Conserved Sequence Required for Normal Splicing, but some Occur in Sequences not Normally Required for Splicing

Many genes naturally undergo alternative forms of RNA splicing. In addition, mutations can sometimes produce an aberrant form of RNA splicing which is pathogenic. Sometimes this results in the sequences of whole exons being excluded from the mature RNA (exon skipping) or retention of whole introns. On other occasions, the abnormal splicing pattern may exclude part of a normal exon or result in new exonic sequences. Point mutations which alter a conserved sequence that is normally required for RNA splicing are comparatively common. Occasionally, however, aberrant splicing of a gene can be induced by mutation of other sequence elements which resemble splice donor or splice acceptor sequences but which are not normally involved in splicing (cryptic splice sites).

Mutations Which Alter Important Splice Site Signals

Often such mutations occur at the essentially invariant GT and AG dinucleotides located respectively at the start of an intron (*splice donor*) or at its end (*splice acceptor*). Flanking these important signals, however, are other conserved sequence elements which, if mutated, can also cause aberrant splicing. Mutations which alter such sequences can have different consequences:

- *Failure of splicing causing intron retention.* This can occasionally result, for example, when an intron is small and the neighbouring sequence lacks alternative legitimate splice sites or *cryptic splice sites* (sequences which resemble the consensus splice site sequences but which are not normally used by the splicing apparatus). The introduction of intronic sequence into the coding sequence of a mature mRNA will, at the very least, introduce additional amino acids and may cause a frameshift.
- *Exon skipping.* The splicing apparatus uses an *alternative* legitimate splice site. Mutation of a splice donor sequence results in skipping of the upstream exon while mutation of the splice acceptor sequence results in skipping of the downstream exon. Often, the exclusion of an exon has a profound effect on gene expression: it may result in a frameshift, an unstable RNA transcript, or a nonfunctional polypeptide because of a loss of a critical group of amino acids.

In addition to the above, the *branch site* used in splicing may be mutated leading to defective splicing.

Mutations of Sequences Which are not Normally Important for RNA Splicing

Cryptic splice sites coincidentally resemble the sequences of authentic splice sites but are not normally used in splicing, unless a mutation alters the sequence so that the splicing apparatus now recognizes it as a normal splice site. The use of an intronic cryptic splice site will introduce new amino acids, while using an exonic cryptic splice site will result in a deletion of coding DNA. The former is a cautionary reminder that apparently silent mutations may yet be pathogenic. Note that in some cases mutations which occur within exons but not at cryptic splice sites can also induce skipping of that exon.

Mutations That Introduce a Premature Termination Codon Usually Result in Unstable mRNA but other Outcomes are Possible

Several different classes of mutation can introduce a premature termination codon (chain terminating mutations). Nonsense mutations produce a premature termination codon simply by substituting a normal codon with a stop codon. Frameshifting insertions and deletions usually also introduce a premature termination codon not too far downstream of the mutation site.

This happens because there is no selection pressure to avoid stop codons in the other translational reading frames and so given established nucleotide frequencies, at least one stop codon is usually encountered within a stretch of 100 nucleotides downstream of the mutation site.

A variety of splice site mutations too can introduce a premature termination codon e.g. by skipping of a single exon containing a number of nucleotides that cannot be divided by 3.

There are several possible consequences for gene expression for chain-terminating mutations:

- *Unstable mRNA.* This is by far the most frequent consequence. A mRNA carrying a premature codon is usually rapidly degraded *in vivo* by a form of RNA surveillance known as nonsense-mediated mRNA decay (Hentze and Kulozik, 1999; Culbertson, 1999). This can avoid the potentially lethal consequences of producing a truncated polypeptide which could interfere with vital cell functions.
- *Truncated polypeptide.* The production of a polypeptide truncated at the C terminus is a very rare outcome *in vivo* (the well-known protein truncation test which assays for mutations introducing a premature termination codon is carried out using an *in vitro* transcription-translation system). Nevertheless, some truncated polypeptides are produced *in vivo.* The effect on gene expression may be difficult to predict and will depend among other things on the extent of the truncation, the stability of the polypeptide product and its ability to interfere with expression of normal alleles.
- *Exon skipping.* Some nonsense mutations appear to induce skipping of constitutive exons *in vivo.* For example, a nonsense mutation in the middle of exon 51 of the *FBN1* fibrillin gene (corresponding to the C terminus of the protein) causes that exon to be skipped (Dietz *et al.*, 1993). As a result of exon skipping the abnormally spliced mRNA uses the normal stop codon and escapes nonsense-mediated mRNA decay unlike any full length mRNA which may be produced from the pre-mRNA. The abnormally spliced *FBN1* mRNA accumulates and is translated to give a dominant negative protein lacking C-terminal sequences.

The Pathogenic Potential of Repeated Sequences

The human genome, like other mammalian genomes, has a very high proportion of DNA sequences that are repeated. Tandem repeats in coding DNA include very short nucleotide repeats, moderately sized repeats and very large repeats that can include whole genes. Depending on the degree of sequence homology between the repeats, tandem repeats are liable to a variety of different genetic mechanisms causing sequence exchange between the repeats.

Often such sequence exchanges result in changes in the number of tandem repeats. A reduction in repeat number can often result in a pathogenic deletion, but expansion by sequence duplication can be pathogenic too. Certain chromosomal regions, notably the subtelomeric and pericentromeric regions, harbour large tracts of duplicated DNA and instability of such regions can predispose to disease (Eichler, 1998). Interspersed repeats can also cause pathogenic mutations by a different variety of mechanisms.

Slipped Strand Mispairing of Short Tandem Repeats Predisposes to Pathogenic Deletions and Frameshifting Insertions

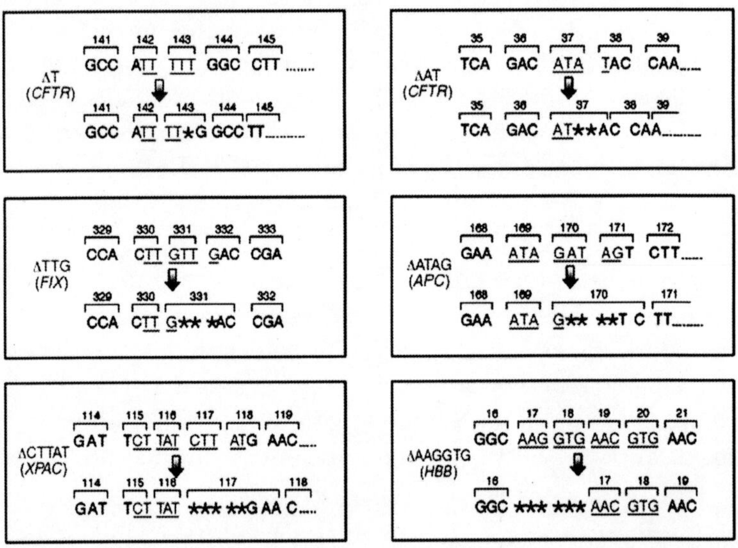

Figure: *Short tandem repeats are deletion/insertion hotspots.*

Insertions and deletions in coding DNA are rare because they usually introduce a translational frameshift. However, occasionally, a series of tandem repeats of a small number of nucleotides occurs by chance in the coding sequence for a polypeptide. Such repeats, like classical microsatellite loci, are comparatively prone to mutation by slipped strand mispairing. As a result, the copy number of tandem repeats is liable to fluctuate, introducing a deletion or an insertion of one or more repeat units. If the mutation occurs in polypeptide-encoding DNA, a resulting deletion will often have a profound effect on gene expression. Frameshifting deletions will normally result in abolition of gene expression. Even if the deletion does not produce a frameshift, deletions of one or more amino acids can still be pathogenic. Small frameshifting insertions will also be expected to lead to loss of gene expression and often the insertion is a tandem repeat of sequences flanking it.

However, nonframeshifting insertions would often not be expected to be pathogenic, unless the insertion occurs in a critically important region, destabilising an essential structure or impeding gene function in some way. Note that large triplet repeat expansions can lead to disease by mechanisms that are not understood at present.

Rapid Large-scale Expansion of Intragenic Triplet Repeats can Cause a Variety of Diseases but the Mutational Mechanism is Not well Understood

Sometimes microsatellites within or in the immediate vicinity of a gene can expand to considerable lengths and affect gene expression, causing disease. In some cases, a modestly expanded repeat which causes disease may be perfectly stable and be propagated without change in size through several generations. For example, triplet repeat expansion leading to long polyalanine tracts in the *HOXD13* gene cause a form of synpolydactyly, probably as a result of unequal crossover (Warren, 1997), but the expanded repeat is stable (Akarsu *et al.,* 1996). In other cases, however, the expanded triplet repeat is *unstable* and the discovery that human disease can be caused by large-scale expansion of highly unstable trinucleotide repeats was quite unexpected. Studies in other organisms had not revealed precedents for such a phenomenon, but the list of human examples is now considerable. In addition to unstable triplet repeat expansion, the majority of disease alleles at the cystatin B gene which cause progressive myoclonus epilepsy involve expansions of a 12 nucleotide repeat (CCCCGCCCCGCG; Lalioti *et al.,* 1998).

Tandem trinucleotide repeats are not infrequent in the human genome. Although there are 64 possible trinucleotide sequences, when allowance is made for cyclic permutations $(CAG)_n = (AGC)_n = (GCA)_n$ and reading from either strand 2 $(CAG)_n$ on one strand = 52 $(CTG)_n$ on the other], there are only 10 different trinucleotide repeats. Most of these are known as usefully polymorphic microsatellite markers but, in addition, certain repeats show anomalous behaviour which can cause abnormal gene expression. In each case, repeats below a certain length are stable in mitosis and meiosis while, above a certain threshold length, the repeats become extremely unstable.

These unstable repeats are virtually never transmitted unchanged from parent to child. Both expansions and contractions can occur, but there is a bias towards expansion. The average size change often depends on the sex of the transmitting parent, as well as the length of the repeat.

The ten possible trinucleotide repeats. Both DNA strands are shown. All other trinculeotide repeats are cyclic permutations of one or another of these.

- *Genes which show modest expansions of* $(CAG)_n$ *repeats within the coding sequence.* Typically, the stable and nonpathological alleles have 10–30 repeats, while unstable pathological alleles have modest expansions, often in the range of 40–200 repeats. Transcription and translation of the gene are not affected by the expansion. The resulting protein product shows a gain of function: its long polyglutamine tract causes it to aggregate within certain cells and kill them.
- *Genes which show very large expansions of a noncoding repeat.* For some genes, various types of triplet repeat (e.g. CGG, CCG, CTG, GAA) found in the promoter, the untranslated regions or intronic sequences can undergo very large expansions in such a way as to inhibit gene expression, causing loss of function. Typically, the stable and nonpathological alleles have 5–50 repeats, while unstable pathological alleles have several hundreds or thousands of copies.

Intergenerational changes are normally reported as parent-child comparisons of blood lymphocyte DNA. There is little information about when in gametogenesis, fertilization or embryogenesis the changes arise. Limited studies of sperm show that highly expanded *DM* and *FRAXA* (fragile-X syndrome) repeats are not transmitted by affected males, although modest expansions can be. The largest expansions in Huntington disease (which, however, are small compared with large *FRAXA* or *DM* expansions) are seen in sperm, consistent with the observation that the severest cases inherit the disease from their father. At least for *FRAXA, DM* and Kennedy disease, the expanded repeats are mitotically unstable, so that a blood sample shows a smear of heterogeneous expanded repeats sizes.

However, *in vitro*, even large repeats are stable. Thus, whatever the mechanism, it is not operative in all cells. The basis of the unstable expansions is very largely unknown and this type of mutagenic event has not been identified thus far in genetically tractable organisms such as *E. coli*, yeast or *Drosophila.* There is also evidence that the unstable expansion mechanism may not have a parallel in some other mammals such as mice. Human transgenes containing long trinucleotide repeats show virtually no instability after being propagated through several generations in transgenic mice whereas the same sequences may show a 100% probability of expansion when transmitted in the human germline (Djian, 1998). Investigations have also suggested that arrays of triplet repeats may be able to form alternative DNA structures, such as DNA hairpins, triplex DNA and quadruplex DNA (Sinden, 1999) but their significance if formed *in vivo* is unknown. The repeats have also been envisaged as possible protein-binding sites and protein-binding at the RNA level has also been envisaged to contribute to pathogenesis in some cases, notably in myotonic dystrophy (Philips *et al.*, 1998).

Slipped strand mispairing is likely to be a component of the expansion mechanism, given the observation that interrupted repeats appear to be stable and only homogeneous repeats are unstable. For example, in spinocerebellar ataxia type 1, 123/126 normal sized CAG repeats were interrupted by one or two CAT triplets, while 30/30 expanded alleles contained no interruption. One problem with all these mispairing mechanisms is that they should result in contractions as well as expansions and this is not seen. Instead, after a certain threshold size, there appears to be a clear bias towards continued expansion of the size of the repeat unit array. Because understanding of trinucleotide repeats is progressing very rapidly at the time of writing, the reader is advised to consult a recent review for more information.

Tandemly Repeated and Clustered gene Families may be Prone to Pathogenic Unequal Crossover and Gene Conversion-like Events

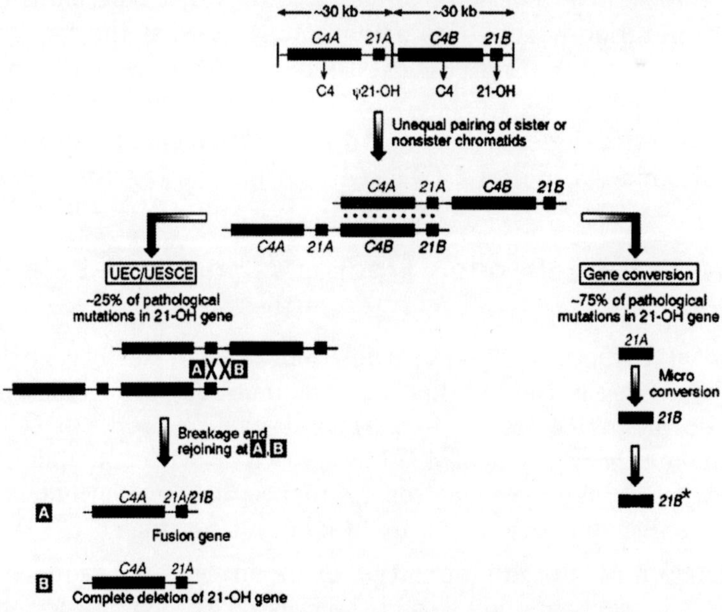

Figure: *Almost all 21-hydroxylase gene mutations are due to sequence exchange with a closely related pseudogene.*

Many human and mammalian gene clusters contain nonfunctional pseudogenes which may be closely related to functional gene members. Interlocus sequence exchanges between pseudogenes and functional genes can result in disease by removing or altering some or all of the sequence of a functional gene. For example, unequal crossover (or unequal sister chromatid exchange) between a functional gene and a related pseudogene can result in deletion of the functional gene or the formation of fusion genes containing a segment derived from the pseudogene. Alternatively, the pseudogene can act as a donor sequence in gene conversion events and introduce deleterious mutations into the functional gene.

The classical example of pathogenesis due to gene- pseudogene exchanges is steroid 21-hydroxylase deficiency, where over 95% of pathogenic mutations arise as a result of sequence exchanges between the functional 21-hydroxylase gene, *CYP21B*, and a very closely related pseudogene, *CYP21A*. The two genes occur on tandemly repeated DNA segments approximately 30 kb long which also contain other duplicated genes, notably the complement C4 genes, *C4A* and *C4B*.

Large pathogenic deletions uniformly result in removal of about 30 kb of DNA, corresponding to one repeat unit length, and analysis of *de novo* 21-hydroxylase deficiency mutations has provided strong evidence for pathogenic deletions arising as a result of meiotic unequal crossover (Sinnott *et al.,* 1990). Virtually all of the 75% of pathogenic point mutations are copied from deleterious mutations in the pseudogene, suggesting a gene conversion mechanism. Analysis of one such mutation which arose *de novo* suggests that the conversion tract is a maximum of 390 bp (Collier *et al.,* 1993). Gene conversion events are also found in the duplicated C4 genes, both of which are normally expressed. A likely priming event for conversions in the *CYP21-C4* gene cluster is unequal pairing of chromatids so that a *CYP21A-C4A* unit pairs with a *CYP21B-C4B* unit.

Interspersed Repeats often Predispose to Large Deletions and Duplications

In several cases, the endpoints of deletions are marked by very short direct repeats. For example, the breakpoints in numerous pathogenic deletions in mtDNA occur at perfect or almost perfect short direct repeats. Of these, the most common is a deletion of 4977 bp which has been found in multiple patients with Kearns-Sayre syndrome, an encephalomyopathy characterised by external ophthalmoplegia, ptosis, ataxia and cataract.

The deletion results in elimination of the intervening sequence between two perfect 13 bp repeats and loss of the sequence of one of the repeats. The mitochondrial genome is recombination-deficient and Shoffner *et al.* 1989 have postulated that such deletions arise by a replication slippage mechanism, similar to that occurring at short tandem repeats. Partial duplications of the mitochondrial genome are also distinctive features of certain diseases, notably Kearns-Sayre syndrome.

The ends of the duplicated sequences, like those of the common deletions, are often marked by short direct repeats, and the mechanisms of duplication and deletion appear to be closely related (Poulton and Holt, 1994).

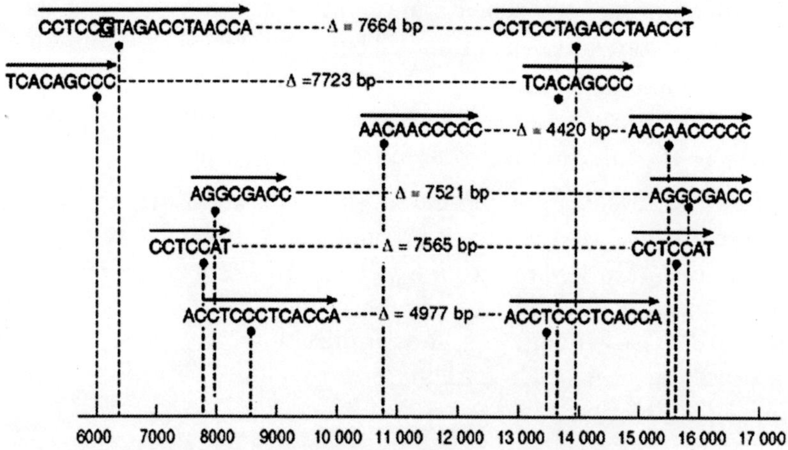

Figure: *Short direct repeats mark the endpoints of many pathogenic deletions in the mitochondrial genome.*

The Alu Repeat as a Recombination Hotspot

Some large-scale deletions and insertions may be generated by pairing of nonallelic interspersed repeats, followed by breakage and rejoining of chromatid fragments. For example, the *Alu* repeat occurs approximately once every 4 kb and mispairing between such repeats has been suggested to be a frequent cause of deletions and duplications.

Some large genes have many internal *Alu* sequences in their introns or untranslated sequences, making them liable to frequent internal deletions and duplications.

For example, the 45-kb low density lipoprotein receptor gene has a relatively high density of *Alu* repeats (approximately one every 1.6 kb). A very high frequency of pathogenic deletions in this gene are likely to involve an *Alu* repeat, usually at both endpoints, and occasional pathogenic intragenic duplications also involve *Alu* repeats (Hobbs *et al.*, 1990).

Such observations have suggested a general role for *Alu* sequences in promoting recombination and recombination-like events. Initial gene duplications in the evolution of clustered multigene families may often have involved an unequal crossover event between *Alu* repeats or other dispersed repetitive elements. It should be noted, however, that some *Alu*-rich genes do not appear to be loci for frequent *Alu*-mediated recombination.

Pathogenic Inversions can be Produced by Intrachromatid Recombination Between Inverted Repeats

Occasionally, clustered inverted repeats with a high degree of sequence identity may be located within or close to a gene. The high degree of sequence similarity between inverted repeats may predispose to pairing of the repeats by a mechanism that involves a chromatid bending back upon itself. Subsequent chromatid breakage at the mispaired repeats and rejoining can then result in an inversion, in much the same way as the natural mechanism used for the production of some immunoglobulin ê light chains.

The classic example of pathogenic inversions is a mutation which accounts for more than 40% of cases of severe hemophilia A. Intron 22 of the factor VIII gene, *F8*, contains a CpG island from which two internal genes are transcribed: *F8A* in the opposite direction to the host gene *F8*, and *F8B* in the same direction as *F8*. *F8A* belongs to a gene family with two other closely related members located several hundred kilobases upstream of *F8* gene and transcribed in the opposite direction to *F8A*.

As a result, the region between the *F8A* gene and the other two members is susceptible to inversions - the *F8A* gene can pair with either of the other two members on the same chromatid, and subsequent chromatid breakage and rejoining in the region of the paired repeats results in an inversion which disrupts the factor VIII gene.

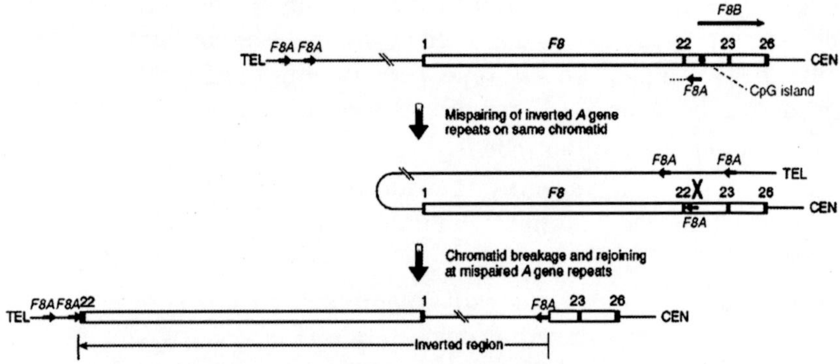

Inversions disrupting the factor VIII gene result from intrachromatid recombination between inverted repeats.

DNA Sequence Transposition is not Uncommon and can Cause Disease

A proportion of moderately and highly repeated interspersed elements are capable of transposition via an RNA intermediate. Defective gene expression due to DNA transposition is comparatively rare and represents only a small component of molecular pathology.

However, several examples have been recorded of genetic deficiency due to insertional inactivation by retrotransposons. For example, in one study, hemophilia A was found to arise in two out of 140 unrelated patients as a result of a *de novo* insertion of a LINE-1 (*Kpn*) repeat into an exon of the factor VIII gene (Kazazian *et al.,* 1988). Other instances are known of insertional inactivation by an actively transposing *Alu* element, as in a case of neurofibromatosis type 1 (Wallace *et al.,* 1991). Additionally, a number of other examples have been recorded of pathogenesis due to intragenic insertion of undefined DNA sequences.

DNA Repair

- DNA in cells suffers a wide range of damage:
- purine bases are lost by spontaneous fission of the base-sugar link;
- cytosines, and occasionally adenines, spontaneously deaminate to produce uracil and hypoxanthine respectively;
- many chemicals, for example alkylating agents, form adducts with DNA bases;
- reactive oxygen species in the cell attack purine and pyrimidine rings
- ultraviolet light causes adjacent thymines to form a stable chemical dimer;
- mistakes in DNA replication result in incorporation of a mismatched base;
- ionising radiation causes single- or double-strand breaks;
- mistakes in replication or recombination leave strand breaks in DNA.

All these lesions must be repaired if the cell is to survive. The importance of effective DNA repair systems is highlighted by the severe diseases affecting people with deficient repair systems.

DNA Repair Usually Involves Cutting out and Resynthesising a Whole area of DNA Surrounding the Damage

To cope with all these forms of damage, cells must be capable of several different types of DNA repair. DNA repair seldom involves simply undoing the change that caused the damage. Almost always a stretch of DNA containing the damaged nucleotide(s) is excised and the gap filled by resynthesis. There are at least five main types of DNA repair in human cells:

- *Direct repair* reverses the DNA damage. A specific enzyme is able to dealkylate O^6-alkyl guanine directly. In bacteria thymine dimers can be removed in a photoreactivation reaction that depends on visible light and an enzyme, photolyase. Mammals possess enzymes related to photolyase, but use them for a quite different purpose, to control their circadian clock (Van der Horst *et al.,* 1999).
- *Base excision repair (BER)* uses glycosidase enzymes to remove abnormal bases. An endonuclease, AP endonuclease, cuts the sugar-phosphate backbone at the position of the missing base. A few nucleotides of the DNA strand are stripped back by exonucleases, the gap is filled by

resynthesis, and the remaining nick is sealed by DNA ligase III. The same process is used to repair spontaneous depurination. Interestingly, no human diseases caused by defective BER are known. Maybe any such defect would be lethal, since BER corrects much the commonest type of DNA damage.

- *Nucleotide excision repair (NER)* removes thymine dimers and large chemical adducts. Defects in NER cause the autosomal recessive disease xeroderma pigmentosum (Lambert *et al.*, 1998). Seven complementation groups, XPA-XPG, have been defined by cell fusion studies. XP patients are exceedingly sensitive to UV light. Sun-exposed skin develops thousands of freckles, many of which progress to skin cancer.
- *Post-replication repair* is required to correct double-strand breaks (Haber, 1999). The usual mechanism is a gene conversion-like process (recombinational repair), where a single strand from the homologous chromosome invades the damaged DNA helix. Alternatively, broken ends are rejoined regardless of their sequence, a desperate measure that is likely to cause mutations.

  The eukaryotic machinery for recombination repair is less well defined than the excision repair systems. Human genes involved in this pathway include *NBS* (mutated in Nijmegen breakage syndrome, MIM 251260), *BLM* (mutated in Bloom syndrome, MIM 210900) and the *BRCA2* and maybe *BRCA1* breast cancer susceptibility genes.
- *Mismatch repair* corrects mismatched base pairs caused by mistakes in DNA replication. Cells deficient in mismatch repair have mutation rates 100–1000 times higher than normal, with a particular tendency to replication slippage in homopolymeric runs. In humans the mechanism involves at least five proteins and defects cause hereditary nonpolyposis colon cancer.

All these systems, except for direct repair, require exo and endonucleases, helicases, polymerases and ligases, usually acting in multiprotein complexes that have some components in common. Sorting out the individual pathways has been greatly aided by the very strong conservation of repair mechanisms across the whole spectrum of life. Not only the reaction mechanisms but also the protein structures and even gene sequences are often conserved from *E. coli* to man.

A downside of the conservation is a confusing gene nomenclature, referring sometimes to human diseases (*XPA* etc.), sometimes to yeast mutants (*RAD* genes) and sometimes to mammalian cell complementation systems (ERCC = excision repair cross-complementing) - for example *XPD*, *ERCC2* and *RAD3* are the same gene in man, mouse and yeast. Generally eukaryotes have multiple proteins corresponding to each single protein in *E. coli*, so that, for example, nucleotide excision repair requires six proteins in *E. coli* but at least 30 in mammals.

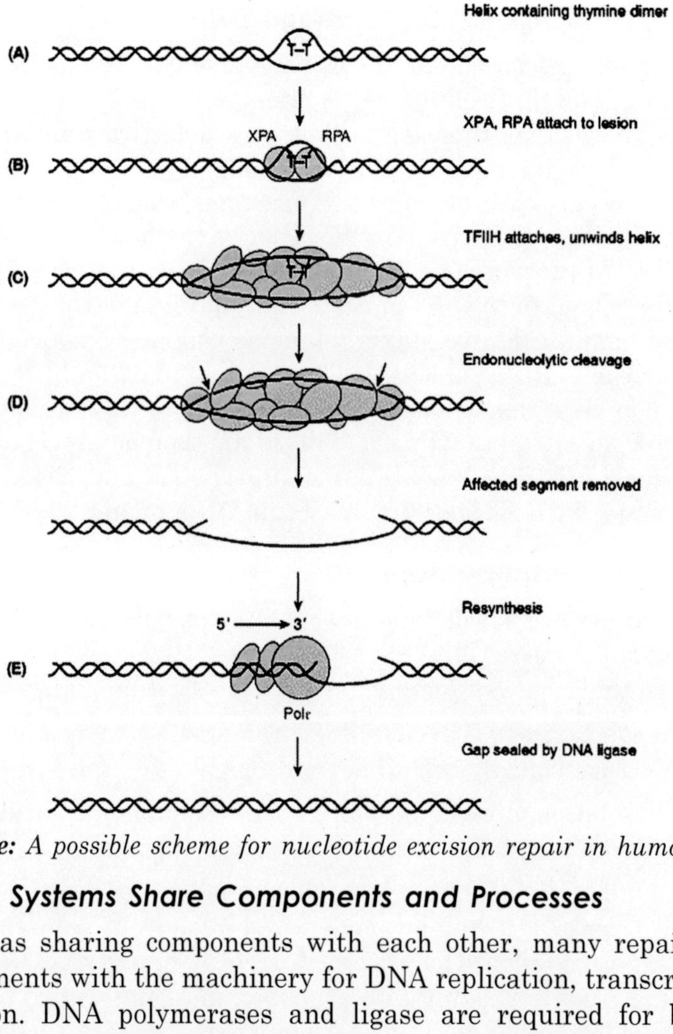

Figure: *A possible scheme for nucleotide excision repair in humans.*

DNA Repair Systems Share Components and Processes

As well as sharing components with each other, many repair systems share components with the machinery for DNA replication, transcription and recombination. DNA polymerases and ligase are required for both DNA replication and resynthesis after excision of a defect. The recombination machinery is involved in double-strand break repair. The link with transcription is particularly intriguing (Lehmann, 1995).

The general transcription factor TFIIH is a multiprotein complex that includes the XPB and XPD proteins. TFIIH exists in two forms. One form is concerned with general transcription and the other with repair, probably specifically repair of transcriptionally active DNA. This system is deficient in two rare diseases, Cockayne syndrome (MIM 216400) and trichothiodystrophy (MIM 601675). Clinically and in cell biology, CS and TTD both overlap XP, and in some cases the same genes are responsible, but CS and TTD patients have developmental defects that presumably reflect defective transcription, and they do not have the cancer susceptibility of XP patients.

Hypersensitivity to Agents that Damage DNA

Many human diseases that involve hypersensitivity to DNA-damaging agents, or a high level of cellular DNA damage, are not caused by defects in the DNA repair systems themselves, but by a defective cellular response to DNA damage. Normal cells react to DNA damage by stalling progress through the cell cycle at a checkpoint until the damage has been repaired, or triggering apoptosis if the damage is irrepairable. Part of the machinery for doing this involves the ATM protein. The role of ATM. Briefly, it senses DNA damage and relays the signal to the p53 protein, the 'guardian of the genome'. People with no functional ATM have ataxia telangiectasia (MIM 208900; Lambert *et al.*, 1998). Their cells are hypersensitive to radiation, and they have chromosomal instability and a high risk of malignancy, but the DNA repair machinery itself is intact. Fanconi anemia (MIM 227650) is another heterogeneous group of diseases (at least five complementation groups) marked by defective responses to DNA damage, without specific defects in DNA repair.

Physical and Transcript Mapping

A wide variety of physical mapping strategies have been used to analyse the DNA of complex eukaryotic genomes. In the following sections, an arbitrary distinction is made between two classes of physical mapping:

- low resolution physical mapping - the smallest map unit that can be resolved is typically one to several megabases of DNA;
- high resolution physical mapping - the resolution is typically very high, from hundreds of kilobases to a single nucleotide.

Since mammalian DNA has only a very small percentage of coding DNA (~3% in the case of the human genome), a variety of transcript mapping methods have been developed for selectively identifying and studying transcribed sequences, the most important and, arguably, most interesting component of the genome. 1Low resolution physical mapping

Somatic cell Hybrid Panels can Permit Chromosomal Localisation of any Human DNA Sequence

Under certain experimental conditions, cultured cells from different species can be induced to fuse together, thereby generating somatic cell hybrids. In human genetic mapping, hybrid cells are typically constructed by fusing human cells and mouse or hamster cells. The initial fusion products are described as heterokaryons because the cells contain both a human and a rodent nucleus. Eventually, heterokaryons proceed to mitosis, and the two nuclear envelopes dissolve. Thereafter, the human and rodent chromosomes are brought together in a single nucleus. The hybrid cells are unstable. For reasons that remain unknown, most human chromosomes fail to replicate in subsequent rounds of cell division and are lost.

This gives rise, eventually, to a variety of more-or-less stable hybrid cell lines, each with the full set of rodent chromosomes plus a few human chromosomes. The loss of the human chromosomes occurs essentially at random, but can be controlled by selection. The human chromosomes in somatic cell hybrids can conveniently be identified by PCR screening with sets of chromosome-specific primers (Abbott and Povey, 1991).

By collecting hybrid cell lines with different human chromosome contents it is possible to generate a hybrid cell panel that can be used to map any human DNA sequence to a specific chromosome. To do this, each of the hybrid cell lines is tested for the presence of the human sequence of interest.

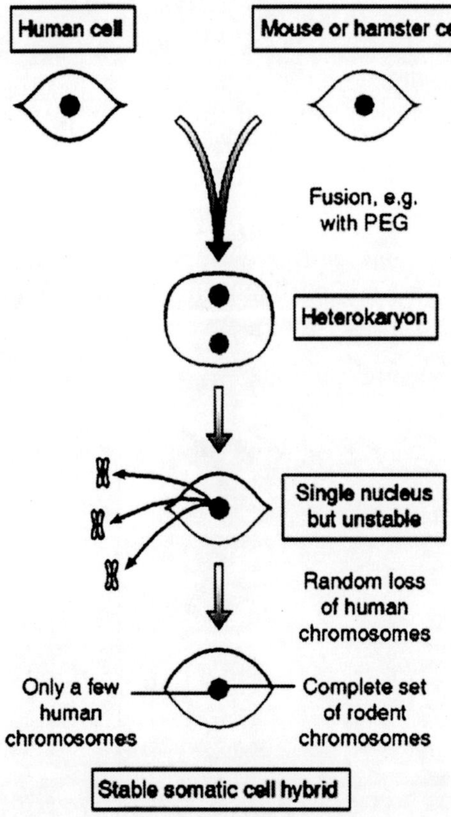

Figure: *Fusion of cells from different species can result in stable somatic cell hybrids.*

A PCR assay can be used with primers specific for that sequence (this is a form of STS content mapping) or the relevant DNA sequence can be labelled and used as a hybridisation probe. Both of these methods require that the human sequence should be distinguishable from any homologous rodent sequence. Somatic cell hybrids could also be used to establish which chromosome encodes a human enzyme or other gene product by biochemical typing of the hybrids. Whichever human chromosome contains the sequence of interest, it will be present in all the cell lines that are positive for the test and absent from all cell lines that are negative for the test.

Monochromosomal hybrids generated by microcell fusion allow direct and rapid chromosomal localisation of human DNA clones. Traditional somatic cell hybrids generally contain several human chromosomes. Microcell fusion (Fournier and Ruddle, 1977) can be used to generate monochromosomal hybrids that contain only a single human chromosome.

The donor human cells are subjected to prolonged mitotic arrest by continued exposure to an inhibitor of mitotic spindle formation, such as colcemid. This causes the chromosome content of a cell to become partitioned into discrete subnuclear packets (micronuclei). Centrifugation in the presence of cytochalasin B (a mitotic spindle inhibitor), results in the formation of microcells. Microcells consist of a single micronucleus with a thin rim of cytoplasm, surrounded by an intact plasma membrane.

In much the same way as for normal donor cells, the microcells can be fused with recipient rodent cells to generate hybrids (microcell-mediated chromosome transfer). Some microcell hybrids contain a few donor chromosomes, but the simplest contain a single donor chromosome.

Subchromosomal Mapping

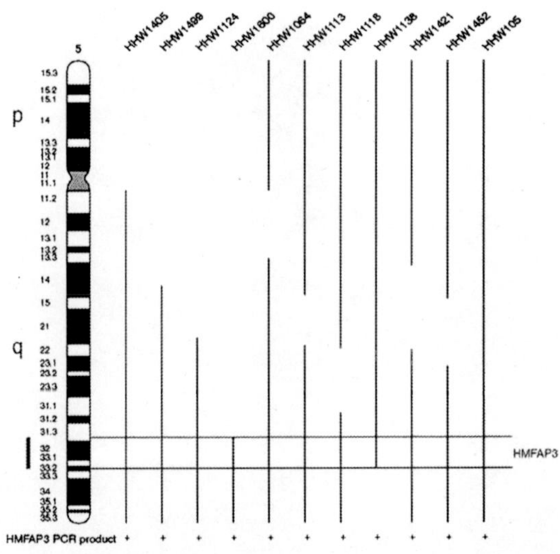

Figure: *Subchromosomal localisation can be achieved by mapping against a panel of hybrid cells containing translocation or deletion chromosomes.*

Monochromosomal human-rodent hybrid cell lines have been established for each of the human chromosomes (Cuthbert *et al.,* 1995). If the sequence of a particular human DNA is known, a PCR assay can be established and used to type a complete monochromosomal hybrid panel, thereby establishing the chromosomal location quickly. Alternatively, if no DNA sequence is available, a human fragment of interest can still be labelled and used as a hybridisation probe against a Southern blot of genomic DNA samples from a monochromosomal hybrid panel. Conventional somatic cell hybrids are a relatively crude tool for physical mapping. More refined mapping is possible using hybrids that contain only part of a particular human chromosome. Translocation hybrids and deletion hybrids are made using donor human cells that have a chromosomal translocation or deletion. To be useful, the hybrids must lack the normal homolog of the chromosome of interest. Such hybrids can be used for subchromosomal mapping of a human sequence-tagged site or biochemical marker. They are especially useful for defining the sequences removed by microdeletions, by segregating the deletion-carrying chromosome away from its normal homolog.

Human DNA Clones can be Physically Mapped to any Desired Resolution

A more general approach to producing tools for physical mapping involves artificially breaking human chromosomes and transferring the fragments into rodent cells.

Chromosome-mediated Gene Transfer

One of the first techniques to use this approach was chromosome-mediated gene transfer (CMGT). Fragments of purified mitotic chromosomes from a donor, such as a human fibroblast, are coprecipitated with calcium phosphate on to the surface of a recipient rodent cell line in monolayer culture. Human chromosome fragments enter the recipient cells, such as mouse fibroblasts, and integrate into the chromosomes, resulting in stable transformation. As a result, hybrids can be established that retain segments of human DNA (transgenomes) of a size that is useful for mapping (usually in the range of 1–50 Mb). However, the transgenomes are prone to frequent re-arrangements, so CMGT is more suited to functional assays of complex loci than as a mapping tool.

Irradiation Fusion gGene Transfer

The most valuable hybrids for gene mapping are radiation hybrids. Donor cells are subjected to a lethal dose of radiation which fragments their chromosomes. The average size of a fragment is a function of the dose of radiation. After irradiation the donor cells are fused with recipient cells of a different species. A selection system is used to pick out recipient cells that have taken up some of the donor chromosome fragments. These cells are useful for mapping insofar as they have taken up a random set of other chromosome fragments from the donor, as well as the selected fragment. Stably incorporated donor fragments are either integrated into rodent chromosomes or are assembled into novel human minichromosomes formed around fragments containing a functional centromere.

Although this procedure was first proposed by Goss and Harris in 1975, it was not used seriously until 1990, when hybrids were constructed using irradiated mono-chromosomal hybrid cells as donors (Cox *et al.*, 1990).

When a set of DNA markers from the human chromosome is assayed in a panel of such radiation hybrids, the patterns of cross-reactivity can be used to construct a map. The frequency of breakage between two markers can be defined by a value θ, analogous to the recombination frequency in meiotic mapping. θ varies from 0 (the two markers are never separated) to 1.0 (the two markers are always broken apart).

As in meiotic mapping, θ underestimates the distance between markers that are far apart on the same chromosome, in this case because a cell can take up two markers on separate fragments. A more accurate estimate is provided by a mapping function, $D = -\ln(1 - \theta)$, which is analogous to the Haldane mapping function used in meiotic linkage analysis. D is measured in centirays (cR). D is dependent on the dosage of radiation, so it is referenced against the number of rads. For example, a distance of 1 cR_{8000} between two markers represents a 1% frequency of breakage between them after exposure to 8000 rad of X-rays.

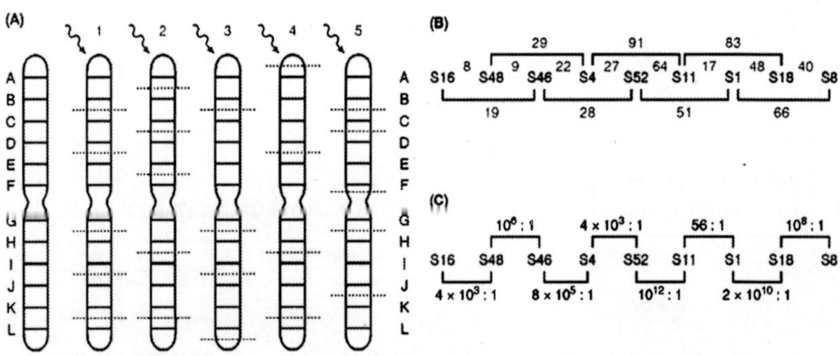

Figure: *Constructing radiation hybrid maps.*

Radiation hybrids derived from monochromosomal hybrid donor cells have been superseded by whole-genome radiation hybrids where the donor is an irradiated normal human diploid cell. The first such panel consisted of 199 hybrids made by fusing an irradiated 46,XY human fibroblast cell line to TK$^-$ hamster cells (Walter *et al.*, 1994). Gyapay *et al.* (1996) used 404 microsatellite markers of known location to show that this hybrid panel could generate accurate maps, and then used it to map 374 unmapped ESTs. A subset of 93 of the hybrids has been made widely available as the Genebridge 4 panel. The

93 hybrids average 32% retention of any particular human sequence, with an average fragment size of 25 Mb. Laboratories can map any unknown STS by scoring the 93 Genebridge hybrids and comparing the pattern with patterns of previously mapped markers held on a central server. This has turned into an extremely powerful and convenient tool for physically mapping any STS or EST. A second human-hamster panel, Stanford G3, was made using a higher dose of radiation, so that the average human fragment size is smaller. The 83 hybrids in G3 average 16% retention of the human genome, with an average fragment size of 2.4 Mb. Thus G3 can be used for finer mapping.

Chromosomal in Situ Hybridisation has been Revolutionised by Fluorescence in Situ Hybridisation Techniques

Chromosomal *in situ* hybridisation typically uses an air-dried microscope slide preparation of metaphase chromosomes, in which the chromosomal DNA has been denatured by exposure to formamide before hybridisation with a suitable probe. Isotopes such as ^{32}P are not suitable for probe labelling because their high energy causes scattering of the signal. Instead, isotopes with weaker emission of beta radiation have been used. Owing to the high background noise when using such isotopes, however, the genuine signal is not easily identified and complex statistical analysis is required to differentiate genuine signal from background noise. More recently, the sensitivity and resolution of *in situ* hybridisation has been increased significantly by the development of fluorescence *in situ* hybridisation (FISH).

In this technique, the DNA probe is either labelled directly by incorporation of a fluorescent labelled nucleotide precursor, or indirectly by incorporation of a nucleotide containing a reporter molecule (such as biotin or digoxigenin) which after incorporation into the DNA is then bound by fluorescently labelled affinity molecule. To increase the intensity of the hybridisation signal, large DNA probes are preferred, usually cosmid clones containing around 40 kb of insert. Because such large sequences will contain many interspersed repetitive DNA sequences, it is necessary to use chromosome *in situ* suppression hybridisation (Lichter *et al.,* 1990). Essentially, this is a form of competition hybridisation: before the main hybridisation, the probe is mixed with a large excess of unlabelled total genomic DNA and denatured, thereby saturating the repetitive elements in the probe, so that they no longer mask the signal generated by the unique sequences.

FISH has the advantage of providing rapid results which can be scored conveniently by eye using a fluorescence microscope. In metaphase spreads, positive signals show as double spots, corresponding to probe hybridised to both sister chromatids. Using sophisticated image processing equipment and reporter-binding molecules carrying different fluorophores, it is possible to map and order several DNA clones simultaneously.

The maximum resolution of conventional FISH on metaphase chromosomes is several megabases. The use of the more extended prometaphase chromosomes can permit 1 Mb resolution but, because of problems with chromatin folding, two differentially labelled probe signals may appear to be side-by-side, unless they are separated by distances greater than 2 Mb. Recently, however, new variations have been developed, permitting very high resolution. There are numerous applications for chromosome FISH but a special one is chromosome painting which has permitted a form of molecular karyotyping.

Flow Cytometry Permits Sorting of Individual Chromosomes and the Construction of Chromosome-specific DNA Libraries

Human chromosomes show considerable variation in size and DNA content. In addition, the base composition can vary considerably: chromosomes that are known to be gene-rich or gene-poor have a comparatively high % (G + C) or % (A + T) respectively. For example, chromosomes 21 and 22 are similar in size and in DNA content, but chromosome 21 is gene-poor and has a relatively high % (A + T), whereas chromosome 22 is gene-rich and has a relatively high % (G + C). As a result of the size differences and differences in base composition, human chromosomes can be separated by flow sorting (also called flow cytometry). In this technique, chromosome preparations are stained with a DNA-binding dye which can fluoresce in a laser beam.

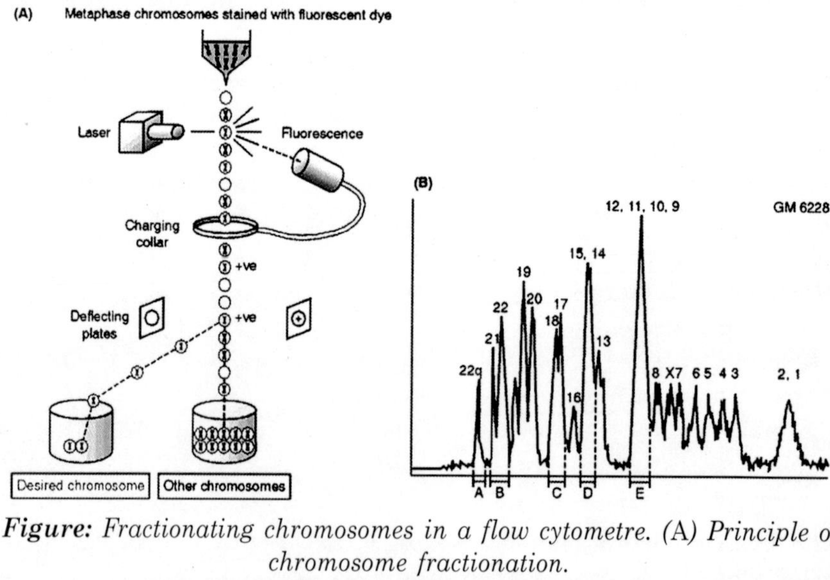

Figure: *Fractionating chromosomes in a flow cytometre. (A) Principle of chromosome fractionation.*

The amount of fluorescence exhibited by a given chromosome is proportional to the amount of dye bound, which in turn is largely proportional to the amount of DNA, and hence the size of the chromosome. Additionally, certain dyes, such as chromomycin A_3, show preferential binding to GC-rich sequences, while others, such as Hoechst 33258, bind preferentially to AT-rich regions. Because individual chromosomes can bind different amounts of fluorescent dyes, they can be fractionated in a flow cytometre. A stream of droplets containing stained chromosomes is passed through a finely focused laser beam at a rate of about 1000–2000 chromosomes per second, and the fluorescence of individual chromosomes present in individual droplets is monitored by a photomultiplier. The intensity of fluorescence of the different chromosomes can then be recorded.

By using sort windows to define a range of fluorescence intensities corresponding to a particular type of chromosome, droplets containing the desired chromosome can be given an electric charge and deflected from the main stream on to a separate collecting grid.

The resulting flow karyogram may show good separation of some human chromosomes; others such as chromosomes 9–12 are not well separated. However, monochromosomal somatic cell hybrids can be used instead, and often result in good separation of the human chromosome. As a result, panels of chromosome-specific DNA representing each of the 24 human chromosomes have been obtained. These can be used to obtain a chromosomal localisation for any human DNA fragment: the fragment is labelled and hybridised to dot-blots of DNA from the 24 chromosomes ('flow-blots'), or a PCR assay can be used, if appropriate. Additionally, purified DNA from a specific type of chromosome can be amplified by cell-based or PCR-based DNA cloning to generate chromosome-specific DNA libraries (Davies et al., 1981).

Chromosome libraries constructed using cosmid vectors have been particularly useful because YACs from a specific chromosome can be related quickly to cosmid clones from the same region. In addition, it has been possible in some cases to construct human chromosome-specific YAC libraries.

Clone Gridding and Increasing Automation Have Facilitated Large-scale Physical Mapping Efforts

Using current technologies, physical mapping of complex genomes is arduous and continues to be hampered by various practical difficulties:

- Chimeric DNA clones. Cloning artifacts can mean that cells contain two noncontiguous pieces of DNA from the desired genome. This may occur as a result of *co-ligation* of two different restriction fragments, internal deletion and *cotransformation*. Yeast cells that have been transformed by YACs and used to propagate have often been much more recombination proficient than typical *E. coli* hosts, and so YACs are often chimeric.
- Repetitive DNA. Arranging clones in a physical order means relying on some clone-specific features. DNA sequences that are repeated can make the task of ordering clones a difficult one, and may cause problems in sequencing DNA.

- Unclonable sequences. Some DNA sequences, including classes of repetitive DNA that are not well represented in bacterial and yeast cells, may not easily be propagated in such cells. A variety of different genomic libraries may need to be screened (including those constructed with vectors containing a low copy number origin of replication such as BAC and P1 libraries).

A major requirement for efficient physical mapping of complex genomes is the need for gridded DNA libraries. Following transformation of the host cells, individual bacterial or yeast colonies are picked into individual wells of microtiter dishes and stored in numerous multiwell dishes. As a result, each cell clone can be assigned an identifying grid coordinate, comprising dish number, row number and column number. When libraries need to be screened by hybridisation with a labelled DNA probe, it is again most efficient to have previously prepared high density gridded colony filters in which the colonies have been spotted on to the appropriate membrane in a gridded fashion with a uniform number of rows and columns.

This simplifies identification of hybridisation-positive colonies. More importantly, it facilitates integration of the results from numerous independent research groups. This is possible because of increasing automation. Robots have been designed to pick and spot colonies, and numerous copies of reference libraries can be distributed, either as cell cultures or, more conveniently, as sets of colony filters and sent to researchers throughout the world. Hybridisation data can then be sent back from individual laboratories to the distributing centres and the fixed geometry of the library coordinates simplifies the task of integrating the information received.

The Human Genome Project is one of several projects to benefit from such technologies, and sets of hybridisation filters containing gridded colonies for total genome YAC libraries, chromosome-specific cosmid libraries and other libraries have been distributed extensively. In addition, pools of yeast colonies have been distributed in agarose plugs to permit PCR typing for a sequence of interest. Again, this is done in a hierarchical coordinate manner to simplify the screening of libraries.

Clone Contigs can be Assembled Rapidly over Large Fractions of a Genome by Random clone Fingerprinting

Chromosome walking is a highly directional and location-restricted procedure for generating clone contigs: a great deal of effort is invested in generating a contig from a fixed starting point on a specific chromosome, often as a prelude to positional cloning. In order to build clone contigs spanning large amounts of a chromosome, or even a whole genome, a more general, labour-efficient mapping approach is needed. To meet this need, a variety of clone fingerprinting techniques have been used to type clones at random and then integrate the information in order to identify clone contigs over large regions of a genome.

For example, this approach has been used to great effect in assembling YAC contigs over significant amounts of the human genome. Clone fingerprinting techniques should be simple and rapid, and the methods described below have been widely used.

Repetitive DNA Fingerprinting

Mammalian DNA has a very high component of highly repeated, interspersed DNA sequences such as the human *Alu* and LINE-1 (*Kpn*) repeats (occurring on average about once every 3 kb and 10 kb respectively).

Because the spacing between such interspersed repetitive DNA elements will vary in different genomic regions, clones which contain overlapping DNA sequences may be identified by similar patterns following hybridisation of repetitive DNA probes to Southern blots of the clone DNA digested with a suitable restriction endonuclease.

For example, a whole genome clone fingerprinting approach has been applied to mapping the human genome largely on the basis of repetitive DNA fingerprinting of YACs (Bellanne-Chantelot *et al.*, 1992).

5

Chromosomes

A chromosome is an organized building of DNA and protein that is found in cells. It is a single piece of coiled DNA containing many genes, regulatory elements and other nucleotide sequences. Chromosomes also contain DNA-bound proteins, which serve to package the DNA and control its functions.

Chromosomes vary widely between different organisms. The DNA molecule may be circular or linear, and can be composed of 10,000 to 1,000,000,000 nucleotides in a long chain. Typically eukaryotic cells (cells with nuclei) have large linear chromosomes and prokaryotic cells (cells without defined nuclei) have smaller circular chromosomes, although there are many exceptions to this rule. Furthermore, cells may contain more than one type of chromosome; for example, mitochondria in most eukaryotes and chloroplasts in plants have their own small chromosomes.

In eukaryotes, nuclear chromosomes are packaged by proteins into a condensed structure called chromatin. This allows the very long DNA molecules to fit into the cell nucleus. The structure of chromosomes and chromatin varies through the cell cycle. Chromosomes are the essential unit for cellular division and must be replicated, divided, and passed successfully to their daughter cells so as to ensure the genetic diversity and survival of their progeny.

Chromosomes may exist as either duplicated or unduplicated—unduplicated chromosomes are single linear strands, whereas duplicated chromosomes (copied during synthesis phase) contain two copies joined by a centromere. Compaction of the duplicated chromosomes during mitosis and meiosis results in the classic four-arm structure (pictured to the right). Chromosomal recombination plays a vital role in genetic diversity. If these structures are manipulated incorrectly, through processes known as chromosomal instability and translocation, the cell may undergo mitotic catastrophe and die, or it may aberrantly evade apoptosis leading to the progression of cancer.

In practice "chromosome" is a rather loosely defined term. In prokaryotes and viruses, the term genophore is more appropriate when no chromatin is present. However, a large body of work uses the term chromosome regardless of chromatin content. In prokaryotes DNA is usually arranged as a circle, which is tightly coiled in on itself, sometimes accompanied by one or more smaller, circular DNA molecules called plasmids. These small circular genomes are also found in mitochondria and chloroplasts, reflecting their bacterial origins. The simplest genophores are found in viruses: these DNA or RNA molecules are short linear or circular genophores that often lack structural proteins.

History

Chromosomes as Vectors of Heredity

In a series of experiments, Theodor Boveri gave the definitive demonstration that chromosomes are the vectors of heredity. His two principles were based upon the *continuity* of chromosomes and the *individuality* of chromosomes. It is the second of these principles that was so original. Boveri was able to test the proposal put forward by Wilhelm Roux, that each chromosome carries a different genetic load, and showed that Roux was right. Upon the rediscovery of Mendel, Boveri was able to point out the connection between the rules of inheritance and the behaviour of the chromosomes. It is interesting to see that Boveri influenced two generations of American cytologists: Edmund Beecher Wilson, Walter Sutton and Theophilus Painter were all influenced by Boveri (Wilson and Painter actually worked with him). In his famous textbook *The Cell*, Wilson linked Boveri and Sutton together by the Boveri-Sutton theory. Mayr remarks that the theory was hotly contested by some famous geneticists: William Bateson, Wilhelm Johannsen, Richard Goldschmidt and T.H. Morgan, all of a rather dogmatic turn-of-mind. Eventually complete proof came from chromosome maps in Morgan's own lab.

Chromosomes in Eukaryotes

Eukaryotes (cells with nuclei such as those found in plants, yeast, and animals) possess multiple large linear chromosomes contained in the cell's nucleus. Each chromosome has one centromere, with one or two arms projecting from the centromere, although, under most circumstances, these arms are not visible as such. In addition, most eukaryotes have a small circular mitochondrial genome, and some eukaryotes may have additional small circular or linear cytoplasmic chromosomes. In the nuclear chromosomes of eukaryotes, the uncondensed DNA exists in a semi-ordered structure, where it is wrapped around histones (structural proteins), forming a composite material called chromatin.

Eukaryotic Chromosome Fine Structure

Eukaryotic chromosome fine structure refers to the structure of sequences for eukaryotic chromosomes. Some fine sequences are included in more than one class, so the classification listed is not intended to be completely separate.

Chromosomal Characteristics

Some sequences are required for a properly functioning chromosome:

- Centromere: Used during cell division as the attachment point for the spindle fibres.
- Telomere: Used to maintain chromosomal integrity by capping off the ends of the linear chromosomes. This region is a microsatellite, but its function is more specific than a simple tandem repeat.

Structural Sequences

Other sequences are used in replication or during interphase with the physical structure of the chromosome.

- Ori, or Origin: Origins of replication.
- MAR: Matrix attachment regions, where the DNA attaches to the nuclear matrix.

Protein-coding Genes

Regions of the genome with protein-coding genes include several elements:

- Enhancer regions (normally up to a few thousand basepairs upstream of transcription).
- Promoter regions (normally less than a couple of hundred basepairs upstream of transcription) include elements such as the TATA and CAAT boxes, GC elements, an initiator, *etc.*
- Exons are the part of the transcript that will eventually be transported to the cytoplasm for translation. When discussing gene with alternate splicing, an exon is a portion of the transcript that could be translated, given the correct splicing conditions. The exons can be divided into three parts
  - o The coding region is the portion of the mRNA that will eventually be translated.
  - o Upstream untranslated region (5' UTR) can serve several functions, including mRNA transport, and initiation of translation (including,

portions of the Kozak sequence). They are never translated into the protein (excepting various mutations).

- o The 3' region downstream from the stop codon is separated into two parts:
  - * 3' UTR is never translated, but serves to add mRNA stability. It is also the attachment site for the poly-A tail. The poly-A tail is used in the initiation of translation and also seems to have an effect on the long-term stability (aging) of the mRNA.
  - * An unnamed region after the poly-A tail, but before the actual site for transcription termination, is spliced off during transcription, and so does not become part of the 3' UTR. Its function, if any, is unknown.

- Introns are intervening sequences between the exons that are never translated. Some sequences inside introns function as miRNA, and there are even some cases of small genes residing completely within the intron of a large gene.

  For some genes (such as the antibody genes), internal control regions are found inside introns. These situations, however, are treated as exceptions.

Genes that are Used as RNA

Many regions of the DNA are transcribed with RNA as the functional form:

- rRNA: Ribosomal RNA are used in the ribosome.
- tRNA: Transfer RNA are used in the translation process by bringing amino acids to the ribosome.
- snRNA: Small nuclear RNA are used in spliceosomes to help the processing of pre-mRNA.
- gRNA: Guide RNA are used in RNA editing.
- miRNA: Micro RNA are small (approximately 24 nucleotides) that are used in gene silencing.
- snoRNA: Small nucleolar RNA are used to help process and construct the ribosome.

Other RNAs are transcribed and not translated, but have undiscovered functions.

Repeated Sequences

Repeated sequences are of two basic types: unique sequences that are repeated in one area; and repeated sequences that are interspersed throughout the genome.

Satellites

Satellites are unique sequences that are repeated in tandem in one area. Depending on the length of the repeat, they are classified as either:

- *Minisatellite:* Short repeats of nucleotides.
- *Microsatellite:* Very short repeats of nucleotides. Some trinucleotide repeats are found in coding regions. Most are found in non-coding regions. Their function is unknown, if they have any specific function. They are used as molecular markers and in DNA fingerprinting.

Interspersed Sequences

Interspersed sequences are tandem repeats, with sequences that are found interspersed across the genome. They can be classified based on the length of the repeat as:

- *SINE:* Short interspersed sequences. The repeats are normally a few hundred base pairs in length. These sequences constitute about 13% of the human genome with the specific *Alu* sequence accounting for 5%.
- *LINE:* Long interspersed sequences. The repeats are normally several thousand base pairs in length. These sequences constitute about 21% of the human genome.

Retrotransposons

Retrotransposons are sequences in the DNA that are the result of retrotransposition of RNA. LINEs and SINEs are examples where the sequences are repeats, but there are non-repeated sequences that can also be retrotransposons.

Other Sequences

Typical eukaryotic chromosomes contain much more DNA than is classified in the categories above. The DNA may be used as spacing, or have other as-yet-unknown function. Or, they may simply be random sequences of no consequence.

Chromatin

Chromatin is the complex of DNA and protein found in the eukaryotic nucleus, which packages chromosomes. The structure of chromatin varies significantly between different stages of the cell cycle, according to the requirements of the DNA.

Interphase Chromatin

During interphase (the period of the cell cycle where the cell is not dividing), two types of chromatin can be distinguished:

- Euchromatin, which consists of DNA that is active, e.g., being expressed as protein.
- Heterochromatin, which consists of mostly inactive DNA. It seems to serve structural purposes during the chromosomal stages. Heterochromatin can be further distinguished into two types:
    - o *Constitutive heterochromatin*, which is never expressed. It is located around the centromere and usually contains repetitive sequences.
    - o *Facultative heterochromatin*, which is sometimes expressed.

Individual chromosomes cannot be distinguished at this stage – they appear in the nucleus as a homogeneous tangled mix of DNA and protein.

Metaphase Chromatin and Division

In the early stages of mitosis or meiosis (cell division), the chromatin strands become more and more condensed. They cease to function as accessible genetic material (transcription stops) and become a compact transportable form. This compact form makes the individual chromosomes visible, and they form the classic four arm structure, a pair of sister chromatids attached to each other at the centromere. The shorter arms are called *p arms* (from the French *petit*, small) and the longer arms are called *q arms* (*q* follows *p* in the Latin alphabet). This is the only natural context in which individual chromosomes are visible with an optical microscope.

During divisions, long microtubules attach to the centromere and the two opposite ends of the cell. The microtubules then pull the chromatids apart, so that each daughter cell inherits one set of chromatids. Once the cells have divided, the chromatids are uncoiled and can function again as chromatin. In spite of their appearance, chromosomes are structurally highly condensed, which enables these giant DNA structures to be contained within a cell nucleus. The self-assembled microtubules form the spindle, which attaches to chromosomes at specialized structures called kinetochores, one of which is present on each sister chromatid. A special DNA base sequence in the region of the kinetochores provides, along with special proteins, longer-lasting attachment in this region.

Mitosis

Mitosis is the process by which a eukaryotic cell separates the chromosomes in its cell nucleus into two identical sets in two nuclei. It is generally followed immediately by cytokinesis, which divides the nuclei, cytoplasm, organelles and cell membrane into two cells containing roughly equal shares of these cellular components. Mitosis and cytokinesis together define the mitotic (M) phase of the cell cycle - the division of the mother cell into two daughter cells, genetically identical to each other and to their parent cell. This accounts for approximately 10% of the cell cycle. Mitosis occurs exclusively in eukaryotic cells, but occurs in different ways in different species. For example, animals undergo an "open" mitosis, where the nuclear envelope breaks down before the chromosomes separate, while fungi such as *Aspergillus nidulans* and *Saccharomyces cerevisiae* (yeast) undergo a "closed" mitosis, where chromosomes divide within an intact cell nucleus. Prokaryotic cells, which lack a nucleus, divide by a process called binary fission.

The process of mitosis is complex and highly regulated. The sequence of events is divided into phases, corresponding to the completion of one set of activities and the start of the next. These stages are interphase, prophase, prometaphase, metaphase, anaphase and telophase. During the process of mitosis the pairs of chromosomes condense and attach to fibres that pull the sister chromatids to opposite sides of the cell. The cell then divides in cytokinesis, to produce two identical daughter cells.

Because cytokinesis usually occurs in conjunction with mitosis, "mitosis" is often used interchangeably with "mitotic phase". However, there are many cells where mitosis and cytokinesis occur separately, forming single cells with multiple nuclei. This occurs most notably among the fungi and slime moulds, but is found in various different groups. Even in animals, cytokinesis and mitosis may occur independently, for instance during certain stages of fruit fly embryonic development. Errors in mitosis can either kill a cell through apoptosis or cause mutations that may lead to cancer.

Overview

The primary result of mitosis is the division of the parent cell's genome into two daughter cells. The genome is composed of a number of chromosomes - complexes of tightly-coiled DNA that contain genetic information vital for proper cell function. Because each resultant daughter cell should be genetically identical to the parent cell, the parent cell must make a copy of each chromosome before mitosis. This occurs during S phase, in interphase, the period that precedes the mitotic phase in the cell cycle where preparation for mitosis occurs.

Each new chromosome now contains two identical copies of itself, called *sister chromatids*, attached together in a specialized region of the chromosome known as the *centromere*. Each sister chromatid is not considered a chromosome in itself, and a chromosome always contains two sister chromatids.

In most eukaryotes, the nuclear envelope that separates the DNA from the cytoplasm disassembles. The chromosomes align themselves in a line spanning the cell. Microtubules, essentially miniature strings, splay out from opposite ends of the cell and shorten, pulling apart the sister chromatids of each chromosome. As a matter of convention, each sister chromatid is now considered a chromosome, so they are renamed to *sister chromosomes*. As the cell elongates, corresponding sister chromosomes are pulled toward opposite ends. A new nuclear envelope forms around the separated sister chromosomes.

As mitosis completes cytokinesis is well underway. In animal cells, the cell pinches inward where the imaginary line used to be (the pinching of the cell membrane to form the two daughter cells is called the cleavage furrow), separating the two developing nuclei. In plant cells, the daughter cells will construct a new dividing cell wall between each other. Eventually, the mother cell will be split in half, giving rise to two daughter cells, each with an equivalent and complete copy of the original genome. Prokaryotic cells undergo a process similar to mitosis called binary fission. However, prokaryotes cannot be properly said to undergo mitosis because they lack a nucleus and only have a single chromosome with no centromere.

Phases of Cell Cycle and Mitosis

Interphase

The mitotic phase is a relatively short period of the cell cycle. It alternates with the much longer *interphase*, where the cell prepares itself for cell division. Interphase is therefore not part of mitosis. Interphase is divided into three phases, G_1 (first gap), S (synthesis), and G_2 (second gap). During all three phases, the cell grows by producing proteins and cytoplasmic organelles. However, chromosomes are replicated only during the S phase. Thus, a cell grows (G_1), continues to grow as it duplicates its chromosomes (S), grows more and prepares for mitosis (G_2), and finally it divides (M) before restarting the cycle.

Preprophase

In plant cells only, prophase is preceded by a pre-prophase stage. In highly vacuolated plant cells, the nucleus has to migrate into the centre of the cell before mitosis can begin. This is achieved through the formation of a phragmosome, a transverse sheet of cytoplasm that bisects the cell along the future plane of cell division. In addition to phragmosome formation, preprophase is characterized by the formation of a ring of microtubules and actin filaments (called preprophase band) underneath the plasma membrane around the equatorial plane of the future mitotic spindle. This band marks the position where the cell will eventually divide.

The cells of higher plants (such as the flowering plants) lack centrioles: with microtubules forming a spindle on the surface of the nucleus and then being organized into a spindle by the chromosomes themselves, after the nuclear membrane breaks down. The preprophase band disappears during nuclear envelope disassembly and spindle formation in prometaphase.

Prophase

Normally, the genetic material in the nucleus is in a loosely bundled coil called chromatin. At the onset of prophase, chromatin condenses together into a highly ordered structure called a chromosome. Since the genetic material has already been duplicated earlier in S phase, the replicated chromosomes have two sister chromatids, bound together at the centromere by the cohesion complex. Chromosomes are visible at high magnification through a light microscope. Close to the nucleus are structures called centrosomes, which are made of a pair of centrioles. The centrosome is the coordinating centre for the cell's microtubules. A cell inherits a single centrosome at cell division, which replicates before a new mitosis begins, giving a pair of centrosomes.

The two centrosomes nucleate microtubules (which may be thought of as cellular ropes or poles) to form the spindle by polymerizing soluble tubulin. Molecular motor proteins then push the centrosomes along these microtubules to opposite sides of the cell. Although centrioles help organize microtubule assembly, they are not essential for the formation of the spindle, since they are absent from plants, and centrosomes are not always used in meiosis.

Prometaphase

The nuclear envelope disassembles and microtubules invade the nuclear space. This is called open mitosis, and it occurs in most multicellular organisms. Fungi and some protists, such as algae or trichomonads, undergo a variation called closed mitosis where the spindle forms inside the nucleus or its microtubules are able to penetrate an intact nuclear envelope.

Each chromosome forms two kinetochores at the centromere, one attached at each chromatid. A kinetochore is a complex protein structure that is analogous to a ring for the microtubule hook; it is the point where microtubules attach themselves to the chromosome. Although the kinetochore structure and function are not fully understood, it is known that it contains some form of molecular motor. When a microtubule connects with the kinetochore, the motor activates, using energy from ATP to "crawl" up the tube toward the originating centrosome. This motor activity, coupled with polymerisation and depolymerisation of microtubules, provides the pulling force necessary to later separate the chromosome's two chromatids

When the spindle grows to sufficient length, *kinetochore microtubules* begin searching for kinetochores to attach to. A number of *nonkinetochore microtubules* find and interact with corresponding nonkinetochore microtubules from the opposite centrosome to form the mitotic spindle. Prometaphase is sometimes considered part of prophase. In the fishing pole analogy, the kinetochore would be the "hook" that catches a sister chromatid or "fish". The centrosome acts as the "reel" that draws in the spindle fibres or "fishing line".

Metaphase

As microtubules find and attach to kinetochores in prometaphase, the centromeres of the chromosomes convene along the *metaphase plate* or *equatorial plane*, an imaginary line that is equidistant from the two centrosome poles. This even alignment is due to the counterbalance of the pulling powers generated by the opposing kinetochores, analogous to a tug-of-war between people of equal strength.

In certain types of cells, chromosomes do not line up at the metaphase plate and instead move back and forth between the poles randomly, only roughly lining up along the midline. Metaphase comes from the Greek *ìåôá* meaning "after." Because proper chromosome separation requires that every kinetochore be attached to a bundle of microtubules (spindle fibres), it is thought that unattached kinetochores generate a signal to prevent premature progression to anaphase without all chromosomes being aligned. The signal creates the *mitotic spindle checkpoint.*

Anaphase

When every kinetochore is attached to a cluster of microtubules and the chromosomes have lined up along the metaphase plate, the cell proceeds to anaphase. Two events then occur; First, the proteins that bind sister chromatids together are cleaved, allowing them to separate. These sister chromatids, which have now become distinct sister chromosomes, are pulled apart by shortening kinetochore microtubules and move toward the respective centrosomes to which they are attached. Next, the nonkinetochore microtubules elongate, pulling the centrosomes (and the set of chromosomes to which they are attached) apart to opposite ends of the cell.

The force that causes the centrosomes to move towards the ends of the cell is still unknown, although there is a theory that suggests that the rapid assembly and breakdown of microtubules may cause this movement. These two stages are sometimes called early and late anaphase. Early anaphase is usually defined as the separation of the sister chromatids, while late anaphase is the elongation of the microtubules and the chromosomes being pulled farther apart. At the end of anaphase, the cell has succeeded in separating identical copies of the genetic material into two distinct populations.

Telophase

Telophase is a reversal of prophase and prometaphase events. It "cleans up" the after effects of mitosis. At telophase, the nonkinetochore microtubules continue to lengthen, elongating the cell even more. Corresponding sister chromosomes attach at opposite ends of the cell. A new nuclear envelope, using fragments of the parent cell's nuclear membrane, forms around each set of separated sister chromosomes. Both sets of chromosomes, now surrounded by new nuclei, unfold back into chromatin. Mitosis is complete, but cell division is not yet complete.

Cytokinesis

Cytokinesis is often mistakenly thought to be the final part of telophase; however, cytokinesis is a separate process that begins at the same time as telophase. Cytokinesis is technically not even a phase of mitosis, but rather a separate process, necessary for completing cell division. In animal cells, a cleavage furrow (pinch) containing a contractile ring develops where the metaphase plate used to be, pinching off the separated nuclei. In both animal and plant cells, cell division is also driven by vesicles derived from the Golgi apparatus, which move along microtubules to the middle of the cell. In plants this structure coalesces into a cell plate at the centre of the phragmoplast and develops into a cell wall, separating the two nuclei. The phragmoplast is a microtubule structure typical for higher plants, whereas some green algae use a phycoplast microtubule array during cytokinesis. Each daughter cell has a complete copy of the genome of its parent cell. The end of cytokinesis marks the end of the M-phase.

Significance

Mitosis is important for the maintenance of the chromosomal set; each cell formed receives chromosomes that are alike in composition and equal in number to the chromosomes of the parent cell. Transcription is generally believed to cease during mitosis, but epigenetic mechanisms such as bookmarking function during this stage of the cell cycle to ensure that the "memory" of which genes were active prior to entry into mitosis are transmitted to the daughter cells.

Consequences of Errors

Although errors in mitosis are rare, the process may go wrong, especially during early cellular divisions in the zygote. Mitotic errors can be especially dangerous to the organism because future offspring from this parent cell will carry the same disorder. In *non-disjunction*, a chromosome may fail to separate during anaphase. One daughter cell will receive both sister chromosomes and the other will receive none. This results in the former cell having three chromosomes containing the same genes (two sisters and a homologue), a condition known as *trisomy*, and the latter cell having only one chromosome (the homologous chromosome), a condition known as *monosomy*. These cells are considered aneuploid, a condition often associated with cancer. Mitosis is a traumatic process. The cell goes through dramatic changes in ultrastructure, its organelles disintegrate and reform in a matter of hours, and chromosomes are jostled constantly by probing microtubules.

Occasionally, chromosomes may become damaged. An arm of the chromosome may be broken and the fragment lost, causing deletion. The fragment may incorrectly reattach to another, non-homologous chromosome, causing translocation. It may reattach to the original chromosome, but in reverse orientation, causing inversion. Or, it may be treated erroneously as a separate chromosome, causing chromosomal duplication. The effect of these genetic abnormalities depends on the specific nature of the error. It may range from no noticeable effect to cancer induction or organism death.

Endomitosis

Endomitosis is a variant of mitosis without nuclear or cellular division, resulting in cells with many copies of the same chromosome occupying a single nucleus. This process may also be referred to as endoreduplication and the cells as endoploid. An example of a cell that goes through endomitosis is the megakaryocyte.

Meiosis

In biology, meiosis is a process of reductional division in which the number of chromosomes per cell is cut in half. In animals, meiosis always results in the formation of gametes, while in other organisms it can give rise to spores. As with mitosis, before meiosis begins, the DNA in the original cell is replicated during S-phase of the cell cycle. Two cell divisions separate the replicated chromosomes into four haploid gametes or spores. Meiosis is essential for sexual reproduction and therefore occurs in all eukaryotes (including single-celled organisms) that reproduce sexually. A few eukaryotes, notably the Bdelloid rotifers, have lost the ability to carry out meiosis and have acquired the ability to reproduce by parthenogenesis. Meiosis does not occur in archaea or bacteria, which reproduce via asexual processes such as binary fission.

During meiosis, the genome of a diploid germ cell, which is composed of long segments of DNA packaged into chromosomes, undergoes DNA replication followed by two rounds of division, resulting in four haploid cells. Each of these cells contains one complete set of chromosomes, or half of the genetic content of the original cell. If meiosis produces gametes, these cells must fuse during fertilization to create a new diploid cell, or zygote before any new growth can occur. Thus, the division mechanism of meiosis is a reciprocal process to the joining of two genomes that occurs at fertilization. Because the chromosomes of each parent undergo homologous recombination during meiosis, each gamete, and thus each zygote, will have a unique genetic *blueprint* encoded in its DNA.

Together, meiosis and fertilization constitute sexuality in the eukaryotes, and generate genetically distinct individuals in populations. In all plants, and in many protists, meiosis results in the formation of haploid cells that can divide vegetatively without undergoing fertilization, referred to as spores. In these groups, gametes are produced by mitosis. Meiosis uses many of the same biochemical mechanisms employed during mitosis to accomplish the redistribution of chromosomes. There are several features unique to meiosis, most importantly the pairing and recombination between homologous chromosomes. Meiosis comes from the root -meio, meaning less.

History

Meiosis was discovered and described for the first time in sea urchin eggs in 1876, by noted German biologist Oscar Hertwig (1849–1922). It was described again in 1883, at the level of chromosomes, by Belgian zoologist Edouard Van Beneden (1846–1910), in *Ascaris* worms' eggs. The significance of meiosis for reproduction and inheritance, however, was described only in 1890 by German biologist August Weismann (1834–1914), who noted that two cell divisions were necessary to transform one diploid cell into four haploid cells if the number of chromosomes had to be maintained. In 1911 the American geneticist Thomas Hunt Morgan (1866–1945) observed crossover in *Drosophila melanogaster* meiosis and provided the first genetic evidence that genes are transmitted on chromosomes.

Evolution

Meiosis is thought to have appeared 1.4 billion years ago. The only supergroup of eukaryotes which does not have meiosis in all organisms is excavata. The other five major supergroups, opisthokonts, amoebozoa, rhizaria, archaeplastida and chromalveolates all seem to have genes for meiosis universally present, even if not always functional. Some excavata species do have meiosis which is consistent with the hypothesis that this group is an ancient, paraphyletic grade. An example of a eukaryotic organism in which meiosis does not exist is euglenoid.

Occurrence of Meiosis in Eukaryotic Life Cycles

Meiosis occurs in eukaryotic life cycles involving sexual reproduction, comprising of the constant cyclical process of meiosis and fertilization. This takes place alongside normal mitotic cell division. In multicellular organisms, there is an intermediary step between the diploid and haploid transition where the organism grows. The organism will then produce the germ cells that continue in the life cycle. The rest of the cells, called somatic cells, function within the organism and will die with it. Cycling meiosis and fertilization events produces a series of transitions back and forth between alternating haploid and diploid states. The organism phase of the life cycle can occur either during the diploid state (*gametic* or *diploid* life cycle), during the haploid state (*zygotic* or *haploid* life cycle), or both (*sporic* or *haplodiploid* life cycle, in which there two distinct organism phases, one during the haploid state and the other during the diploid state).

In this sense, there are three types of life cycles that utilize sexual reproduction, differentiated by the location of the organisms phase(s). In the *gametic life cycle*, of which humans are a part, the species is diploid, grown from a diploid cell called the zygote.

The organism's diploid germ-line stem cells undergo meiosis to create haploid gametes (the spermatozoa for males and ova for females), which fertilize to form the zygote. The diploid zygote undergoes repeated cellular division by mitosis to grow into the organism. Mitosis is a related process to meiosis that creates two cells that are genetically identical to the parent cell. The general principle is that mitosis creates somatic cells and meiosis creates germ cells.

In the *zygotic life cycle* the species is haploid instead, spawned by the proliferation and differentiation of a single haploid cell called the gamete. Two organisms of opposing gender contribute their haploid germ cells to form a diploid zygote. The zygote undergoes meiosis immediately, creating four haploid cells. These cells undergo mitosis to create the organism. Many fungi and many protozoa are members of the zygotic life cycle.

Finally, in the *sporic life cycle*, the living organism alternates between haploid and diploid states. Consequently, this cycle is also known as the alternation of generations. The diploid organism's germ-line cells undergo meiosis to produce spores. The spores proliferate by mitosis, growing into a haploid organism. The haploid organism's germ cells then combine with another haploid organism's cells, creating the zygote. The zygote undergoes repeated mitosis and differentiation to become the diploid organism again. The sporic life cycle can be considered a fusion of the gametic and zygotic life cycles.

Process

Because meiosis is a "one-way" process, it cannot be said to engage in a cell cycle as mitosis does. However, the preparatory steps that lead up to meiosis are identical in pattern and name to the interphase of the mitotic cell cycle. Interphase is divided into three phases:

- Growth 1 (G_1) phase: This is a very active period, where the cell synthesizes its vast array of proteins, including the enzymes and structural proteins it will need for growth. In G_1 stage each of the chromosomes consists of a single (very long) molecule of DNA. In humans, at this point cells are 46 chromosomes, 2N, identical to somatic cells.
- Synthesis (S) phase: The genetic material is replicated: each of its chromosomes duplicates, so that each of the 46 chromosomes forms a second identical sister chromatid. The cell is still considered diploid because it still contains the same number of centromeres. The identical sister chromatids have not yet condensed into the densely packaged

chromosomes visible with the light microscope. This will take place during prophase I in meiosis.

- Growth 2 (G_2) phase: G_2 phase is absent in Meiosis.

Interphase is followed by meiosis I and then meiosis II. Meiosis I consists of separating the pairs of homologous chromosome, each made up of two sister chromatids, into two cells. One entire haploid content of chromosomes is contained in each of the resulting daughter cells; the first meiotic division therefore reduces the ploidy of the original cell by a factor of 2. Meiosis II consists of decoupling each chromosome's sister strands (chromatids), and segregating the individual chromatids into haploid daughter cells.

The two cells resulting from meiosis I divide during meiosis II, creating 4 haploid daughter cells. Meiosis I and II are each divided into prophase, metaphase, anaphase, and telophase stages, similar in purpose to their analogous subphases in the mitotic cell cycle. Therefore, meiosis includes the stages of meiosis I (prophase I, metaphase I, anaphase I, telophase I), and meiosis II (prophase II, metaphase II, anaphase II, telophase II). Meiosis generates genetic diversity in two ways: (1) independent alignment and subsequent separation of homologous chromosome pairs during the first meiotic division allows a random and independent selection of each chromosome segregates into each gamete; and (2) physical exchange of homologous chromosomal regions by homologous recombination during prophase I results in new combinations of DNA within chromosomes.

Phases of Meiosis

Meiosis takes place in several stages.

Meiosis I

Meiosis I separates homologous chromosomes, producing two haploid cells (N chromosomes, 23 in humans), so meiosis I is referred to as a reductional division. A regular diploid human cell contains 46 chromosomes and is considered 2N because it contains 23 pairs of homologous chromosomes. However, after meiosis I, although the cell contains 46 chromatids, it is only considered as being N, with 23 chromosomes. This is because later, in Anaphase I, the sister chromatids will remain together as the spindle fibres pulls the pair toward the pole of the new cell. In meiosis II, an equational division similar to mitosis will occur whereby the sister chromatids are finally split, creating a total of 4 haploid cells (23 chromosomes, N) per daughter cell from the first division.

Prophase I

During prophase I, DNA is exchanged between homologous chromosomes in a process called homologous recombination. This often results in chromosomal crossover. The new combinations of DNA created during crossover are a significant source of genetic variation, and may result in beneficial new combinations of alleles. The paired and replicated chromosomes are called bivalents or tetrads, which have two chromosomes and four chromatids, with one chromosome coming from each parent. At this stage, non-sister chromatids may cross-over at points called chiasmata (plural; singular chiasma).

Leptotene

The first stage of prophase I is the *leptotene* stage, also known as *leptonema*, from Greek words meaning "thin threads". During this stage, individual chromosomes begin to condense into long strands within the nucleus. However the two sister chromatids are still so tightly bound that they are indistinguishable from one another.

Zygotene

The *zygotene* stage, also known as *zygonema*, from Greek words meaning "paired threads", occurs as the chromosomes approximately line up with each other into homologous chromosome pairs. This is called the bouquet stage because of the way the telomeres cluster at one end of the nucleus. At this stage, the synapsis (pairing/coming together) of homologous chromosomes takes place.

Pachytene

The *pachytene* stage, also known as *pachynema*, from Greek words meaning "thick threads", contains the following chromosomal crossover. Nonsister chromatids of homologous chromosomes randomly exchange segments of genetic information over regions of homology. Sex chromosomes, however, are not wholly identical, and only exchange information over a small region of homology.

Exchange takes place at sites where *recombination nodules* (the chiasmata) have formed. The exchange of information between the non-sister chromatids results in a recombination of information; each chromosome has the complete set of information it had before, and there are no gaps formed as a result of the process. Because the chromosomes cannot be distinguished in the synaptonemal complex, the actual act of crossing over is not perceivable through the microscope.

Diplotene

During the *diplotene* stage, also known as *diplonema*, from Greek words meaning "two threads", the synaptonemal complex degrades and homologous chromosomes separate from one another a little. The chromosomes themselves uncoil a bit, allowing some transcription of DNA. However, the homologous chromosomes of each bivalent remain tightly bound at chiasmata, the regions where crossing-over occurred. The chiasmata remain on the chromosomes until they are severed in Anaphase I. In human fetal oogenesis all developing oocytes develop to this stage and stop before birth. This suspended state is referred to as the *dictyotene stage* and remains so until puberty. In males, only spermatogonia (spermatogenesis) exist until meiosis begins at puberty.

Diakinesis

Chromosomes condense further during the *diakinesis* stage, from Greek words meaning "moving through". This is the first point in meiosis where the four parts of the tetrads are actually visible. Sites of crossing over entangle together, effectively overlapping, making chiasmata clearly visible. Other than this observation, the rest of the stage closely resembles prometaphase of mitosis; the nucleoli disappear, the nuclear membrane disintegrates into vesicles, and the meiotic spindle begins to form.

Synchronous Processes

During these stages, two centrosomes, containing a pair of centrioles in animal cells, migrate to the two poles of the cell. These centrosomes, which were duplicated during S-phase, function as microtubule organizing centres nucleating microtubules, which are essentially cellular ropes and poles. The microtubules invade the nuclear region after the nuclear envelope disintegrates, attaching to the chromosomes at the kinetochore.

The kinetochore functions as a motor, pulling the chromosome along the attached microtubule toward the originating centriole, like a train on a track. There are four kinetochores on each tetrad, but the pair of kinetochores on each sister chromatid fuses and functions as a unit during meiosis I.

Microtubules that attach to the kinetochores are known as *kinetochore microtubules*. Other microtubules will interact with microtubules from the opposite centriole: these are called *nonkinetochore microtubules* or *polar microtubules*. A third type of microtubules, the aster microtubules, radiates from the centrosome into the cytoplasm or contacts components of the membrane skeleton.

Metaphase I

Homologous pairs move together along the metaphase plate: As kinetochore microtubules from both centrioles attach to their respective kinetochores, the homologous chromosomes align along an equatorial plane that bisects the spindle, due to continuous counterbalancing forces exerted on the bivalents by the microtubules emanating from the two kinetochores of homologous chromosomes.

The physical basis of the independent assortment of chromosomes is the random orientation of each bivalent along the metaphase plate, with respect to the orientation of the other bivalents along the same equatorial line.

Anaphase I

Kinetochore (bipolar spindles) microtubules shorten, severing the recombination nodules and pulling homologous chromosomes apart. Since each chromosome has only one functional unit of a pair of kinetochores, whole chromosomes are pulled toward opposing poles, forming two haploid sets. Each chromosome still contains a pair of sister chromatids. Nonkinetochore microtubules lengthen, pushing the centrioles farther apart. The cell elongates in preparation for division down the centre.

Telophase I

The last meiotic division effectively ends when the chromosomes arrive at the poles. Each daughter cell now has half the number of chromosomes but each chromosome consists of a pair of chromatids. The microtubules that make up the spindle network disappear, and a new nuclear membrane surrounds each haploid set.

The chromosomes uncoil back into chromatin. Cytokinesis, the pinching of the cell membrane in animal cells or the formation of the cell wall in plant cells, occurs, completing the creation of two daughter cells. Sister chromatids remain attached during telophase I. Cells may enter a period of rest known as interkinesis or interphase II. No DNA replication occurs during this stage.

Meiosis II

Meiosis II is the second part of the meiotic process. Much of the process is similar to mitosis. The end result is production of four haploid cells (23 chromosomes, 1N in humans) from the two haploid cells (23 chromosomes, 1N* each of the chromosomes consisting of two sister chromatids) produced in meiosis I. The four main steps of Meiosis II are: Prophase II, Metaphase II, Anaphase II, and Telophase II. In prophase II we see the disappearance of the nucleoli and the nuclear envelope again as well as the shortening and thickening of the chromatids. Centrioles move to the polar regions and arrange spindle fibres for the second meiotic division. In metaphase II, the centromeres contain two kinetochores that attach to spindle fibres from the centrosomes (centrioles) at each pole. The new equatorial metaphase plate is rotated by 90 degrees when compared to meiosis I, perpendicular to the previous plate.

This is followed by anaphase II, where the centromeres are cleaved, allowing microtubules attached to the kinetochores to pull the sister chromatids apart. The sister chromatids by convention are now called sister chromosomes as they move toward opposing poles.

The process ends with telophase II, which is similar to telophase I, and is marked by uncoiling and lengthening of the chromosomes and the disappearance of the spindle. Nuclear envelopes reform and cleavage or cell wall formation eventually produces a total of four daughter cells, each with a haploid set of chromosomes. Meiosis is now complete and ends up with four new daughter cells.

Significance

Meiosis facilitates stable sexual reproduction. Without the halving of ploidy, or chromosome count, fertilization would result in zygotes that have twice the number of chromosomes as the zygotes from the previous generation. Successive generations would have an exponential increase in chromosome count. In organisms that are normally diploid, polyploidy, the state of having three or more sets of chromosomes, results in extreme developmental abnormalities or lethality. Polyploidy is poorly tolerated in most animal species. Plants, however, regularly produce fertile, viable polyploids. Polyploidy has been implicated as an important mechanism in plant speciation.

Most importantly, recombination and independent assortment of homologous chromosomes allow for a greater diversity of genotypes in the population. This produces genetic variation in gametes that promote genetic and phenotypic variation in a population of offspring.

Nondisjunction

The normal separation of chromosomes in meiosis I or sister chromatids in meiosis II is termed *disjunction*. When the separation is not normal, it is called nondisjunction. This results in the production of gametes which have either too many or too few of a particular chromosome, and is a common mechanism for trisomy or monosomy. Nondisjunction can occur in the meiosis I or meiosis II, phases of cellular reproduction, or during mitosis.

This is a cause of several medical conditions in humans (such as):

- Down Syndrome - trisomy of chromosome 21
- Patau Syndrome - trisomy of chromosome 13
- Edward Syndrome - trisomy of chromosome 18
- Klinefelter Syndrome - extra X chromosomes in males - i.e. XXY, XXXY, XXXXY
- Turner Syndrome - lacking of one X chromosome in females - i.e. XO
- Triple X syndrome - an extra X chromosome in females
- XYY Syndrome - an extra Y chromosome in males.

Meiosis in Mammals

In females, meiosis occurs in cells known as oogonia (singular: oogonium). Each oogonium that initiates meiosis will divide twice to form a single oocyte and two polar bodies. However, before these divisions occur, these cells stop at the diplotene stage of meiosis I and lie dormant within a protective shell of somatic cells called the follicle. Follicles begin growth at a steady pace in a process known as folliculogenesis, and a small number enter the menstrual cycle. Menstruated oocytes continue meiosis I and arrest at meiosis II until fertilization. The process of meiosis in females occurs during oogenesis, and differs from the typical meiosis in that it features a long period of meiotic arrest known as the Dictyate stage and lacks the assistance of centrosomes.

In males, meiosis occurs in precursor cells known as spermatogonia that divide twice to become sperm. These cells continuously divide without arrest in the seminiferous tubules of the testicles. Sperm is produced at a steady pace. The process of meiosis in males occurs during spermatogenesis.

In female mammals, meiosis begins immediately after primordial germ cells migrate to the ovary in the embryo, but in the males, meiosis begins years later at the time of puberty. It is retinoic acid, derived from the primitive kidney (mesonephros) that stimulates meiosis in ovarian oogonia. Tissues of the male testis suppress meiosis by degrading retinoic acid, a stimulator of meiosis. This is overcome at puberty when cells within seminiferous tubules called Sertoli cells start making their own retinoic acid. Sensitivity to retinoic acid is also adjusted by proteins called nanos and DAZL. Meiosis involves Spermatocytes.

Chromosomes in Prokaryotes

The prokaryotes – bacteria and archaea – typically have a single circular chromosome, but many variations do exist. Most bacteria have a single circular chromosome that can range in size from only 160,000 base pairs in the endosymbiotic bacterium *Candidatus Carsonella ruddii*, to 12,200,000 base pairs in the soil-dwelling bacterium *Sorangium cellulosum*. Spirochaetes of the genus *Borrelia* are a notable exception to this arrangement, with bacteria such as *Borrelia burgdorferi*, the cause of Lyme disease, containing a single linear chromosome.

Structure in Sequences

Prokaryotic chromosomes have less sequence-based structure than eukaryotes. Bacteria typically have a single point (the origin of replication) from which replication starts, whereas some archaea contain multiple replication origins. The genes in prokaryotes are often organized in operons, and do not usually contain introns, unlike eukaryotes.

DNA Packaging

Prokaryotes do not possess nuclei. Instead, their DNA is organized into a structure called the nucleoid. The nucleoid is a distinct structure and occupies a defined region of the bacterial cell. This structure is, however, dynamic and is maintained and remodeled by the actions of a range of histone-like proteins, which associate with the bacterial chromosome. In archaea, the DNA in chromosomes is even more organized, with the DNA packaged within structures similar to eukaryotic nucleosomes.

Bacterial chromosomes tend to be tethered to the plasma membrane of the bacteria. In molecular biology application, this allows for its isolation from plasmid DNA by centrifugation of lysed bacteria and pelleting of the membranes (and the attached DNA). Prokaryotic chromosomes and plasmids are, like eukaryotic DNA, generally supercoiled. The DNA must first be released into its relaxed state for access for transcription, regulation, and replication. "beads-on-a-string" fibre, the 30nm fibre and the metaphase chromosome.

Number of Chromosomes in Various Organisms

Eukaryotes

These tables give the total number of chromosomes (including sex chromosomes) in a cell nucleus. For example, human cells are diploid and have 22 different types of autosome, each present as two copies, and two sex chromosomes. This gives 46 chromosomes in total. Other organisms have more than two copies of their chromosomes, such as bread wheat, which is *hexaploid* and has six copies of seven different chromosomes – 42 chromosomes in total.

Normal members of a particular eukaryotic species all have the same number of nuclear chromosomes. Other eukaryotic chromosomes, i.e., mitochondrial and plasmid-like small chromosomes, are much more variable in number, and there may be thousands of copies per cell.

Asexually reproducing species have one set of chromosomes, which are the same in all body cells. However, asexual species can be either haploid or diploid. Sexually reproducing species have somatic cells (body cells), which are diploid [2n] having two sets of chromosomes, one from the mother and one from the father. Gametes, reproductive cells, are haploid [n]: They have one set of chromosomes.

Gametes are produced by meiosis of a diploid germ line cell. During meiosis, the matching chromosomes of father and mother can exchange small parts of themselves (crossover), and thus create new chromosomes that are not inherited solely from either parent. When a male and a female gamete merge (fertilization), a new diploid organism is formed.

Some animal and plant species are polyploid [Xn]: They have more than two sets of homologous chromosomes. Plants important in agriculture such as tobacco or wheat are often polyploid, compared to their ancestral species. Wheat has a haploid number of seven chromosomes, still seen in some cultivars as well as the wild progenitors.

The more-common pasta and bread wheats are polyploid, having 28 (tetraploid) and 42 (hexaploid) chromosomes, compared to the 14 (diploid) chromosomes in the wild wheat.

Prokaryotes

Prokaryote species generally have one copy of each major chromosome, but most cells can easily survive with multiple copies. For example, *Buchnera*, a symbiont of aphids has multiple copies of its chromosome, ranging from 10–400 copies per cell.

However, in some large bacteria, such as *Epulopiscium fishelsoni* up to 100,000 copies of the chromosome can be present. Plasmids and plasmid-like small chromosomes are, as in eukaryotes, very variable in copy number. The number of plasmids in the cell is almost entirely determined by the rate of division of the plasmid – fast division causes high copy number, and vice versa.

Karyotype

In general, the karyotype is the characteristic chromosome complement of a eukaryote species. The preparation and study of karyotypes is part of cytogenetics. Although the replication and transcription of DNA is highly standardized in eukaryotes, *the same cannot be said for their karyotypes*, which are often highly variable. There may be variation between species in chromosome number and in detailed organization. In some cases, there is significant variation within species. Often there is:

1. variation between the two sexes
2. variation between the germ-line and soma (between gametes and the rest of the body)
3. variation between members of a population, due to balanced genetic polymorphism
4. geographical variation between races
5. mosaics or otherwise abnormal individuals.

Also, variation in karyotype may occur during development from the fertilised egg. The technique of determining the karyotype is usually called *karyotyping*. Cells can be locked part-way through division (in metaphase) in vitro (in a reaction vial) with colchicine. These cells are then stained, photographed, and arranged into a *karyogram*, with the set of chromosomes arranged, autosomes in order of length, and sex chromosomes (here X/Y) at the end:

Like many sexually reproducing species, humans have special gonosomes (sex chromosomes, in contrast to autosomes). These are XX in females and XY in males.

Historical Note

Investigation into the human karyotype took many years to settle the most basic question. How many chromosomes does a normal diploid human cell contain? In 1912, Hans von Winiwarter reported 47 chromosomes in spermatogonia and 48 in oogonia, concluding an XX/XO sex determination mechanism. Painter in 1922 was not certain whether the diploid number of man is 46 or 48, at first favouring 46. He revised his opinion later from 46 to 48, and he correctly insisted on man's having an XX/XY system. New techniques were needed to definitively solve the problem:

1. Using cells in culture
2. Pretreating cells in a hypotonic solution, which swells them and spreads the chromosomes
3. Arresting mitosis in metaphase by a solution of colchicine
4. Squashing the preparation on the slide forcing the chromosomes into a single plane
5. Cutting up a photomicrograph and arranging the result into an indisputable karyogram.

It took until the mid-1950s for it to become generally accepted that the human karyotype include only 46 chromosomes. Considering the techniques of Winiwarter and Painter, their results were quite remarkable. Chimpanzees (the closest living relatives to modern humans) have 48 chromosomes.

Chromosomal Aberrations

Chromosomal aberrations are disruptions in the normal chromosomal content of a cell, and are a major cause of genetic conditions in humans, such as Down syndrome. Some chromosome abnormalities do not cause disease in carriers, such as translocations, or chromosomal inversions, although they may lead to a higher chance of birthing a child with a chromosome disorder. Abnormal numbers of chromosomes or chromosome sets, aneuploidy, may be lethal or give rise to genetic disorders. Genetic counseling is offered for families that may carry a chromosome rearrangement.

The gain or loss of DNA from chromosomes can lead to a variety of genetic disorders. Human examples include:

- Cri du chat, which is caused by the deletion of part of the short arm of chromosome 5. "Cri du chat" means "cry of the cat" in French, and the condition was so-named because affected babies make high-pitched cries that sound like those of a cat. Affected individuals have wide-set eyes, a small head and jaw, moderate to severe mental health issues, and are very short.

- Down syndrome, usually is caused by an extra copy of chromosome 21 (trisomy 21). Characteristics include decreased muscle tone, stockier build, asymmetrical skull, slanting eyes and mild to moderate developmental disability.
- Edwards syndrome, which is the second-most-common trisomy; Down syndrome is the most common. It is a trisomy of chromosome 18. Symptoms include motor retardation, developmental disability and numerous congenital anomalies causing serious health problems. Ninety percent die in infancy; however, those that live past their first birthday usually are quite healthy thereafter. They have a characteristic clenched hands and overlapping fingers.
- Idic15, abbreviation for Isodicentric 15 on chromosome 15; also called the following names due to various researches, but they all mean the same; IDIC(15), Inverted duplication 15, extra Marker, Inv dup 15, partial tetrasomy 15
- Jacobsen syndrome, also called the terminal 11q deletion disorder. This is a very rare disorder. Those affected have normal intelligence or mild developmental disability, with poor expressive language skills. Most have a bleeding disorder called Paris-Trousseau syndrome.
- Klinefelter's syndrome (XXY). Men with Klinefelter syndrome are usually sterile, and tend to have longer arms and legs and to be taller than their peers. Boys with the syndrome are often shy and quiet, and have a higher incidence of speech delay and dyslexia. During puberty, without testosterone treatment, some of them may develop gynecomastia.
- Patau Syndrome, also called D-Syndrome or trisomy-13. Symptoms are somewhat similar to those of trisomy-18, but they do not have the characteristic hand shape.
- Small supernumerary marker chromosome. This means there is an extra, abnormal chromosome. Features depend on the origin of the extra genetic material. Cat-eye syndrome and isodicentric chromosome 15 syndrome (or Idic15) are both caused by a supernumerary marker chromosome, as is Pallister-Killian syndrome.
- Triple-X syndrome (XXX). XXX girls tend to be tall and thin. They have a higher incidence of dyslexia.
- Turner syndrome (X instead of XX or XY). In Turner syndrome, female sexual characteristics are present but underdeveloped. People with Turner syndrome often have a short stature, low hairline, abnormal eye features and bone development and a "caved-in" appearance to the chest.

- XYY syndrome. XYY boys are usually taller than their siblings. Like XXY boys and XXX girls, they are somewhat more likely to have learning difficulties.
- Wolf-Hirschhorn syndrome, which is caused by partial deletion of the short arm of chromosome 4. It is characterized by severe growth retardation and severe to profound mental health issues.

Chromosomal mutations produce changes in whole chromosomes (more than one gene) or in the number of chromosomes present.

- Deletion – loss of part of a chromosome
- Duplication – extra copies of a part of a chromosome
- Inversion – reverse the direction of a part of a chromosome
- Translocation – part of a chromosome breaks off and attaches to another chromosome.

Most mutations are neutral – have little or no effect. Chromosomal aberrations are the changes in the structure of chromosomes. It has a great role in evolution. A detailed graphical display of all human chromosomes and the diseases annotated at the correct spot may be found at the Oak Ridge National Laboratory.

Human Chromosomes

Chromosomes can be divided into two types—autosomes, and sex chromosomes. Certain genetic traits are linked to your sex, and are passed on through the sex chromosomes. The autosomes contain the rest of the genetic hereditary information. All act in the same way during cell division. Human cells have 23 pairs of large linear nuclear chromosomes, (22 pairs of autosomes and one pair of sex chromosomes) giving a total of 46 per cell. In addition to these, human cells have many hundreds of copies of the mitochondrial genome.

Sequencing of the human genome has provided a great deal of information about each of the chromosomes. Below is a table compiling statistics for the chromosomes, based on the Sanger Institute's human genome information in the Vertebrate Genome Annotation (VEGA) database. Number of genes is an estimate as it is in part based on gene predictions. Total chromosome length is an estimate as well, based on the estimated size of unsequenced heterochromatin regions.

Chromosome	*Genes*	*Total bases*	*Sequenced bases*
1	4,220	247,199,719	224,999,719
2	1,491	242,751,149	237,712,649
3	1,550	199,446,827	194,704,827

4	446	191,263,063	187,297,063
5	609	180,837,866	177,702,766
6	2,281	170,896,993	167,273,993
7	2,135	158,821,424	154,952,424
8	1,106	146,274,826	142,612,826
9	1,920	140,442,298	120,312,298
10	1,793	135,374,737	131,624,737
11	379	134,452,384	131,130,853
12	1,430	132,289,534	130,303,534
13	924	114,127,980	95,559,980
14	1,347	106,360,585	88,290,585
15	921	100,338,915	81,341,915
16	909	88,822,254	78,884,754
17	1,672	78,654,742	77,800,220
18	519	76,117,153	74,656,155
19	1,555	63,806,651	55,785,651
20	1,008	62,435,965	59,505,254
21	578	46,944,323	34,171,998
22	1,092	49,528,953	34,893,953
X (sex chromosome)	1,846	154,913,754	151,058,754
Y (sex chromosome)	454	57,741,652	25,121,652
Total	32,185	3,079,843,747	2,857,698,560

6

Genetic Code

The genetic code is the set of rules by which information encoded in genetic material (DNA or mRNA sequences) is translated into proteins (amino acid sequences) by living cells. The code defines a mapping between tri-nucleotide sequences, called codons, and amino acids. With some exceptions, a triplet codon in a nucleic acid sequence specifies a single amino acid. Because the vast majority of genes are encoded with exactly the same code, this particular code is often referred to as the canonical or standard genetic code, or simply *the* genetic code, though in fact there are many variant codes. For example, protein synthesis in human mitochondria relies on a genetic code that differs from the standard genetic code. Not all genetic information is stored using the genetic code. All organisms' DNA contains regulatory sequences, intergenic segments, and chromosomal structural areas that can contribute greatly to phenotype. Those elements operate under sets of rules that are distinct from the codon-to-amino acid paradigm underlying the genetic code.

Discovery

After the structure of DNA was deciphered by James Watson, Francis Crick, Maurice Wilkins and Rosalind Franklin, serious efforts to understand the nature of the encoding of proteins began. George Gamow postulated that a three-letter code must be employed to encode the 20 standard amino acids used by living cells to encode proteins, because 3 is the smallest integer n such that 4 is at least 20. The fact that codons consist of three DNA bases was first demonstrated in the Crick, Brenner et al. experiment. The first elucidation of a codon was done by Marshall Nirenberg and Heinrich J. Matthaei in 1961 at the National Institutes of Health. They used a cell-free system to translate a poly-uracil RNA sequence (i.e., UUUUU...) and discovered that the polypeptide that they had synthesized consisted of only the amino acid phenylalanine. They thereby deduced that the codon UUU specified the amino acid phenylalanine. This was followed by experiments in the laboratory of Severo Ochoa demonstrating that the poly-adenine RNA sequence (AAAAA...) coded for the polypeptide, poly-lysine. and the poly-cytosine RNA sequence (CCCCC...) coded for the polypeptide, poly-proline. Therefore the codon AAA specified the amino acid lysine, and the codon CCC specified the amino acid proline. Using different copolymers most of the remaining codons were then determined. Extending this work, Nirenberg and Philip Leder revealed the triplet nature of the genetic code and allowed the codons of the standard genetic code to be deciphered. In these experiments various combinations of mRNA were passed through a filter which contained ribosomes, the components of cells that translate RNA into protein. Unique triplets promoted the binding of specific tRNAs to the ribosome. Leder and Nirenberg were able to determine the sequences of 54 out of 64 codons in their experiments.

Subsequent work by Har Gobind Khorana identified the rest of the genetic code. Shortly thereafter, Robert W. Holley determined the structure of transfer RNA (tRNA), the adapter molecule that facilitates the process of translating RNA into protein. This work was based upon earlier studies by Severo Ochoa, who received the Nobel prize in 1959 for his work on the enzymology of RNA synthesis. In 1968, Khorana, Holley and Nirenberg received the Nobel Prize in Physiology or Medicine for their work.

Transfer of Information Via the Genetic Code

The genome of an organism is inscribed in DNA, or in the case of some viruses, RNA. The portion of the genome that codes for a protein or an RNA is referred to as a gene. Those genes that code for proteins are composed of tri-nucleotide units called codons, each coding for a single amino acid. Each nucleotide sub-unit consists of a phosphate, deoxyribose sugar and one of the 4 nitrogenous nucleobases. The purine bases adenine (A) and guanine (G) are larger and consist of two aromatic rings. The pyrimidine bases cytosine (C) and thymine (T) are smaller and consist of only one aromatic ring. In the double-helix configuration, two strands of DNA are joined to each other by hydrogen bonds in an arrangement known as base pairing. These bonds almost always form between an adenine base on one strand and a thymine on the other strand and between a cytosine base on one strand and a guanine base on the other. This means that the number of A and T residues will be the same in a given double helix, as will the number of G and C residues. In RNA, thymine (T) is replaced by uracil (U), and the deoxyribose is substituted by ribose.

Each protein-coding gene is transcribed into a template molecule of the related polymer RNA, known as messenger RNA or mRNA. This, in turn, is translated on the ribosome into an amino acid chain or polypeptide. The process of translation requires transfer RNAs specific for individual amino acids with the amino acids covalently attached to them, guanosine triphosphate as an energy source, and a number of translation factors. tRNAs have anticodons complementary to the codons in mRNA and can be "charged" covalently with amino acids at their 3' terminal CCA ends. Individual tRNAs are charged with specific amino acids by enzymes known as aminoacyl tRNA synthetases, which have high specificity for both their cognate amino acids and tRNAs. The high specificity of these enzymes is a major reason why the fidelity of protein translation is maintained.

There are $4^3 = 64$ different codon combinations possible with a triplet codon of three nucleotides; all 64 codons are assigned for either amino acids or stop signals during translation. If, for example, an RNA sequence, UUUAAACCC is considered and the reading frame starts with the first U (by convention, 5' to 3'), there are three codons, namely, UUU, AAA and CCC, each of which specifies one amino acid. This RNA sequence will be translated into an amino acid sequence, three amino acids long. A comparison may be made with computer science, where the codon is similar to a word, which is the standard "chunk" for handling data (like one amino acid of a protein), and a nucleotide is similar to a bit, in that it is the smallest unit.

Salient Features

Sequence Reading Frame

A codon is defined by the initial nucleotide from which translation starts. For example, the string GGGAAACCC, if read from the first position, contains the codons GGG, AAA and CCC; and, if read from the second position, it contains the codons GGA and AAC; if read starting from the third position, GAA and ACC. Every sequence can thus be read in three reading frames, each of which will produce a different amino acid sequence (in the given example, Gly-Lys-Pro, Gly-Asn, or Glu-Thr, respectively). With double-stranded DNA there are six possible reading frames, three in the forward orientation on one strand and three reverse on the opposite strand. The actual frame in which a protein sequence is translated is defined by a start codon, usually the first AUG codon in the mRNA sequence.

Start/Stop Codons

Translation starts with a chain initiation codon (start codon). Unlike stop codons, the codon alone is not sufficient to begin the process. Nearby sequences (such as the Shine-Dalgarno sequence in *E. coli*) and initiation factors are also required to start translation. The most common start codon is AUG which is read as methionine or, in bacteria, as formylmethionine. Alternative start codons (depending on the organism), include "GUG" or "UUG", which normally code for valine or leucine, respectively. However, when used as a start codon, these alternative start codons are translated as methionine or formylmethionine. The three stop codons have been given names: UAG is *amber*, UGA is *opal* (sometimes also called *umber*), and UAA is *ochre*. "Amber" was named by discoverers Richard Epstein and Charles Steinberg after their friend Harris Bernstein, whose last name means "amber" in German. The other two stop codons were named "ochre" and "opal" in order to keep the "color names" theme. Stop codons are also called "termination" or "nonsense" codons and they signal release of the nascent polypeptide from the ribosome due to binding of release factors in the absence of cognate tRNAs with anticodons complementary to these stop signals.

Effect of Mutations

During the process of DNA replication, errors occasionally occur in the polymerization of the second strand. These errors, called mutations, can have an impact on the phenotype of an organism, especially if they occur within the protein coding sequence of a gene. Error rates are usually very low—1 error in every 10–100 million bases—due to the "proofreading" ability of DNA polymerases. Missense mutations and nonsense mutations are examples of point mutations, which can cause genetic diseases such as sickle-cell disease and thalassemia respectively. Clinically important missense mutations generally change the properties of the coded amino acid residue between being basic, acidic polar or non-polar, whereas nonsense mutations result in a stop codon.

Mutations that disrupt the reading frame sequence by indels (insertions or deletions) of a non-multiple of 3 nucleotide bases are known as frameshift mutations. These mutations usually result in a completely different translation from the original, and are also very likely to cause a stop codon to be read, which truncates the creation of the protein. These mutations may impair the function of the resulting protein, and are thus rare in *in vivo* protein-coding sequences. One reason inheritance of frameshift mutations is rare is that if the protein being translated is essential for growth under the selective pressures the organism faces, absence of a functional protein may cause death before the organism is viable. Frameshift mutations may result in severe genetic diseases such as Tay-Sachs disease. Although most mutations that change protein sequences are harmful or neutral, some mutations have a positive effect on an organism. These mutations may enable the mutant organism to withstand particular environmental stresses better than wild-type organisms, or reproduce more quickly. In these cases a mutation will tend to become more common in a population through natural selection. Viruses that use RNA as their genetic material have rapid mutation rates, which can be an advantage since these viruses will evolve constantly and rapidly, and thus evade the defensive responses of e.g. the human immune system. In large populations of asexually reproducing organisms, for example, E. coli, multiple beneficial mutations may co-occur, causing competition among them, this phenomenon is called clonal interference.

Degeneracy of the Genetic Code

The genetic code has redundancy but no ambiguity. For example, although codons GAA and GAG both specify glutamic acid (redundancy), neither of them specifies any other amino acid (no ambiguity). The codons encoding one amino acid may differ in any of their three positions. For example the amino acid glutamic acid is specified by GAA and GAG codons (difference in the third position), the amino acid leucine is specified by UUA, UUG, CUU, CUC, CUA, CUG codons (difference in the first or third position), while the amino acid serine is specified by UCA, UCG, UCC, UCU, AGU, AGC (difference in the first, second or third position).

A position of a codon is said to be a fourfold degenerate site if any nucleotide at this position specifies the same amino acid. For example, the third position of the glycine codons (GGA, GGG, GGC, GGU) is a fourfold degenerate site, because all nucleotide substitutions at this site are synonymous; i.e., they do not change the amino acid. Only the third positions of some codons may be fourfold degenerate. A position of a codon is said to be a twofold degenerate site if only two of four possible nucleotides at this position specify the same amino acid. For example, the third position of the glutamic acid codons (GAA, GAG) is a twofold degenerate site. In twofold degenerate sites, the equivalent nucleotides are always either two purines (A/G) or two pyrimidines (C/U), so only transversional substitutions (purine to pyrimidine or pyrimidine to purine) in twofold degenerate sites are nonsynonymous. A position of a codon is said to be a non-degenerate site if any mutation at this position results in amino acid substitution.

There is only one threefold degenerate site where changing to three of the four nucleotides may have no effect on the amino acid (depending on what it is changed to), while changing to the fourth possible nucleotide always results in an amino acid substitution. This is the third position of an isoleucine codon: AUU, AUC, or AUA all encode isoleucine, but AUG encodes methionine. In computation this position is often treated as a twofold degenerate site. There are three amino acids encoded by six different codons: serine, leucine, and arginine. Only two amino acids are specified by a single codon. One of these is the amino-acid methionine, specified by the codon AUG, which also specifies the start of translation; the other is tryptophan, specified by the codon UGG. The degeneracy of the genetic code is what accounts for the existence of synonymous mutations. Degeneracy results because a triplet code designates 20 amino acids and a stop codon. Because there are four bases, triplet codons are required to produce at least 21 different codes. For example, if there were two bases per codon, then only 16 amino acids could be coded for ($4^2=16$). Because at least 21 codes are required, then 4^3 gives 64 possible codons, meaning that some degeneracy must exist.

These properties of the genetic code make it more fault-tolerant for point mutations. For example, in theory, fourfold degenerate codons can tolerate any point mutation at the third position, although codon usage bias restricts this in practice in many organisms; twofold degenerate codons can tolerate one out of the three possible point mutations at the third position. Since transition mutations (purine to purine or pyrimidine to pyrimidine mutations) are more likely than transversion (purine to pyrimidine or vice-versa) mutations, the equivalence of purines or that of pyrimidines at twofold degenerate sites adds a further fault-tolerance. A practical consequence of redundancy is that some errors in the genetic code only cause a silent mutation or an error that would not affect the protein because the hydrophilicity or hydrophobicity is maintained by equivalent substitution of amino acids; for example, a codon of NUN (where N = any nucleotide) tends to code for hydrophobic amino acids. NCN yields amino acid residues that are small in size and moderate in hydropathy; NAN encodes average size hydrophilic residues. These tendencies may result from the shared ancestry of the aminoacyl tRNA synthetases related to these codons.

Even so, single point mutations can still cause dysfunctional proteins. For example, a mutated hemoglobin gene causes sickle-cell disease. In the mutant hemoglobin a hydrophilic glutamate (Glu) is substituted by the hydrophobic valine (Val), that is, GAA or GAG becomes GUA or GUG. The substitution of glutamate by valine reduces the solubility of â-globin which causes hemoglobin to form linear polymers linked by the hydrophobic interaction between the valine groups causing sickle-cell deformation of erythrocytes. Sickle-cell disease is generally not caused by a *de novo* mutation. Rather it is selected for in malarial regions (in a way similar to thalassemia), as heterozygous people have some resistance to the malarial *Plasmodium* parasite (heterozygote advantage). These variable codes for amino acids are allowed because of modified bases in the first base of the anticodon of the tRNA, and the base-pair formed is called a wobble base pair. The modified bases include inosine and the Non-Watson-Crick U-G basepair.

Variations to the Standard Genetic Code

While slight variations on the standard code had been predicted earlier, none were discovered until 1979, when researchers studying human mitochondrial genes discovered they used an alternative code. Many slight variants have been discovered since, including various alternative mitochondrial codes, as well as small variants such as *Mycoplasma* translating the codon UGA as tryptophan and *Candida* species translating CUG as a serine rather than a leucine. In bacteria and archaea, GUG and UUG are common start codons. However, in rare cases, certain specific proteins may use alternative initiation (start) codons not normally used by that species. In certain proteins, non-standard amino acids are substituted for standard stop codons, depending upon associated signal sequences in the messenger RNA: UGA can code for selenocysteine and UAG can code for pyrrolysine as discussed in the relevant articles. Selenocysteine is now viewed as the 21st amino acid, and pyrrolysine is viewed as the 22nd.

Notwithstanding these differences, all known codes have strong similarities to each other, and the coding mechanism is the same for all organisms: three-base codons, tRNA, ribosomes, reading the code in the same direction and translating the code three letters at a time into sequences of amino acids.

Expanded Genetic Code

An expanded genetic code refers to an artificially modified genetic code in which one or more specific codons have been allocated to encode an amino acid which is not among the twenty/twentytwo found in nature. In order to understand how expansion of the genetic code is chieved, the translational mechanism catalysed by ribosomes should be considered: the central parts are the transfer RNAs (tRNA) which are used as keys to decode a three nucleotide code on the RNA being translated into its equivalent amino acid. To do this the tRNA recognises a specific codon thanks to a complementary sequence called the anticodon on one of its loops. As a result generally there are one (or more) tRNAs for each codon (or group of degenerate codons). The encoding of a codon to its amino acid is a result of the aminoacyl tRNA synthetase which adds the aminoacyl group to its allocated tRNA. However, the aminoacyl tRNA synthetase often recognises a specific motif irrespective of the anticodon, meaning that if the anticodon were to be mutated the encoding of that amino acid would change to a new codon.

For successful translation of a novel amino acid, the codon to which the amino acid is reassigned must be free or unfavoured and the novel tRNA and synthetase set (called the orthogonal set when including the codon) must not crosstalk with the endogenous tRNA and synthetase sets, while still being functionally compatible with the ribosome and other components of the translation apparatus. The tRNA synthetase pair is taken from a distant organism, generally from a different domain, and the active site of the synthetase is modified to accept the non-natural amino acid. The possibility of reassigning codons was realized by Normanly et al. in 1990 when a viable mutant strain of E. coli read through the amber codon. As a result the amber codon became the choice codon to be assigned a novel amino acid. Later, in the Schultz lab the tRNATyr/tyrosyl-tRNA synthetase (TyrRS) from Methanococcus jannaschii was used to introduce a tyrosine instead of STOP, the default value of the amber codon. As mentioned, this was possible because of the differences between the endogenous bacterial synthases and the orthologous archeal synthase which do not recognise each other.

Directed Evolution

Several methods for selecting the synthetase which accepts only the non natural amino acid have been developed. One of which is by using a combination of positive and negative selection. *Some amino acids which have been added in order to label protein.* This orthologous set can then be mutated and screened through directed evolution to accept a different, even novel, amino acid. Mutations to the plasmid containing the pair can be introduced by error-prone PCR or through degenerate primers for the synthetase's active site. Selection involves multiple rounds of a two-step process, where the plasmid is transferred into cells expressing chloramphenicol acetyl transferase with a premature amber codon. In the presence of toxic chloramphenicol and the non-natural amino acid, the surviving cells will have overridden the amber codon using the orthogonal tRNA aminoacylated with either the standard amino acids or the non-natural one. To remove the former, the plasmid is inserted into cells with a barnase gene (toxic) with a premature amber codon but without the non-natural amino acid, removing all the orthogonal synthases which do not specifically recognize the non-natural amino acid. In addition to the recoding of the tRNA to a different codon, they can be mutated to recognize a four base codon, allowing additional free coding options. The non natural amino acid, as a result, introduces diverse physicochemical and biological properties in order to be used as a tool to explore protein structure and function or to create novel or enhanced protein for practical purposes.

Diversity

The orthogonal pairs of synthase and tRNA which work for one organism may not work for another as the synthase may mis-aminoacylate endogenous tRNAs or the tRNA be mis-aminoacylated itself by an endogenous synthase. As a result the sets created to date different between organisms.

Orthogonal Sets in E. Coli

- tRNA-TyrRS pair from the archaeon *Methanococcus jannaschii*
- tRNA–LysRS pair from the archaeon *Pyrococcus horikoshii*
- tRNA–GluRS pair from *Methanosarcina mazei*
- leucyl-tRNA synthetase from *Methanobacterium hermoautotrophicum* and a mutant leucyl tRNA derived from *Halobacterium* sp.

Orthogonal Sets in Yeast

- tRNA-TyrRS pair from *Escherichia coli*
- tRNA–LeuRS pair from *Escherichia coli*
- tRNA from human and GlnRS from *Escherichia coli.*

Orthogonal Sets in Mammalian Cells

- tRNA-TyrRS pair from *Bacillus stearothermophilus*
- modified tRNA-TrpRS pair from *Bacillus subtilis* trp
- tRNA–LeuRS pair from *Escherichia coli.*

Protein Studies

With an expanded genetic code, the unnatural amino acid can be genetically directed to any chosen site in the protein of interest. The high efficiency and fidelity allows a better control of the placement of the modification compared to modifying the protein post-translationally, which generally will target all amino acids of the same type, such as the thiol group of cysteine and the - amino group of lysine. Also, an expanded genetic code allows modifications to be carried out *in vivo*. The ability to site-specifically direct lab-synthesized chemical moieties into proteins allows many types of studies which would otherwise be extremely difficult.

- Probing Protein Structure and Function: by using amino acids with slightly different size such as o-Methyltyrosine or dansylalanine instead of tyrosine, and by inserting genetically-coded reporter moieties (color-changing and/or spin-active) into selected protein sites, chemical information about the protein's structure and function can be measured.
- Identifying and Regulating Protein Activity: by using photocaged aminoacids, protein function can be "switched" on or off by illuminating the organism.
- Changing the mode of action of a protein: one can start with the gene for a protein which binds a certain sequence of DNA, and, by inserting a chemically active amino acid into the binding site, convert it to a protein which cuts the DNA, rather than binding it.
- Improving immunogenicity and overcoming self-tolerance: by replacing strategically-chosen tyrosines with *p*-nitro phenylalanine, a tolerated self-protein can be made immunogenic.

An example of the possible application for this method is the biomedical where "chemical warheads" can be added to protein which target specific cellular components. Since 2001, 40 non-natural amino acids have been added into protein by creating a unique codon (recoding) and a corresponding transfer-RNA:aminoacyl – tRNA-synthetase pair to encode it with diverse physicochemical and biological properties in order to be used as a tool to exploring protein structure and function or to create novel or enhanced proteins.

Theories on the Origin of the Genetic Code

Despite the minor variations that exist, the genetic code used by all known forms of life is nearly universal. However, there are a huge number of possible genetic codes. If amino acids are randomly associated with triplet codons, there will be 1.5 x 10 possible genetic codes. The question arises: why this code? How did it originate? Phylogenetic analysis of transfer RNA suggests that tRNA molecules evolved before the present set of aminoacyl-tRNA synthetases. Theoretically the genetic code could be completely random (a "frozen accident"), completely non-random (optimal) or a combination of random and nonrandom. There are sufficient data to refute the first possibility. For a start, a quick view on the table of the genetic code already shows a clustering of amino acid assignments. Furthermore, amino acids that share the same biosynthetic pathway tend to have the same first base in their codons, and amino acids with similar physical properties tend to have similar codons. There are four themes running through the many theories that seek to explain the evolution of the genetic code (and hence the origin of these patterns):

- Chemical principles govern specific RNA interaction with amino acids. Aptamer experiments showed that some amino acids have a selective chemical affinity for the base triplets that code for them. Recent experiments show that of the 8 amino acids tested, 6 show some RNA triplet-amino acid association. This has been called the stereochemical code. The stereochemical code could have created an ancient core of assignments. The current complex translation mechanism involving tRNA and associated enzymes may be a later development, and that originally, protein sequences were directly templated on base sequences.
- Biosynthetic expansion. The standard modern genetic code grew from a simpler earlier code through a process of "biosynthetic expansion". Here the idea is that primordial life "discovered" new amino acids (e.g., as by-products of metabolism) and later back-incorporated some of these into the machinery of genetic coding. Although much circumstantial evidence has been found to suggest that fewer different amino acids were used in the past than today, precise and detailed hypotheses about exactly which amino acids entered the code in exactly what order have proved far more controversial.
- Natural selection has led to codon assignments of the genetic code that minimize the effects of mutations. A recent hypothesis suggests that the triplet code was derived from codes that used longer than triplet codons. Longer than triplet decoding has higher degree of codon

redundancy and is more error resistant than the triplet decoding. This feature could allow accurate decoding in the absence of highly complex translational machinery such as the ribosome.

- The biocommunicative approach investigates nucleotide sequences as code, i.e. language-like text, which follows in parallel three (3) kinds of rules: combinatorial (syntactic), context-sensitive (pragmatic) and content-specific (semantic). Natural genome editing from this perspective is competent viral-driven generation and integration of meaningful nucleotide sequences into pre-existing genomic content arrangements and the ability to (re-)combine and (re-)regulate them according to context-dependent (i.e. adaptational) purposes of the host organism.
- Information-theoretic approaches treat the genetic code in terms of an error-prone information channel. The inherent noise in the channel poses the organism with a fundamental question: how to construct a genetic code that can withstand the impact of noise while accurately and efficiently translating information? These "rate-distortion" models suggest that the genetic code originated as a result of the interplay of the three conflicting evolutionary forces: the needs for diverse amino-acids, for error-tolerance and for minimal cost of resources. The code emerges at a coding phase transition when the mapping of codons to amino-acids becomes nonrandom. The emergence of the code is governed by the topology defined by the most probable errors and is related to the map colouring problem.
- Genetic code and Codon populations: In other way, considering the global information level of Universal Genetic Code map, studying the whole Human genome codon populations show that the position of each codon within this table precisely dictates its population. So the Universal Genetic Code Table not only maps codons to amino acids, but serves perhaps as a global checksum matrix at the balance level of the whole genome: frequencies of the 64 codons in the whole human genome scale are a self-similar fractal expansion of the universal genetic code map. This global matrix-like function of Genetic Code table was published in September 2010 by the journal *Interdisciplinary Science* reporting that *Codon populations in single-stranded whole human genome DNA Are fractal and fine-tuned by the Golden Ratio 1.618*. In this paper, Jean-Claude Perez shows that Universal Genetic Code Table kernel governs not only the micro scale (codons i.e amino acids) but the macro scale as well. Particularly, the 6 folding steps of codon populations modeled by the binary divisions of the *Dragon fractal paper folding curve* extend the Chargaff's second rule from the domain of single TCAG nucleotides to the larger domain of codon triplets in the case of the whole Human genome.

Control of Gene Expression in Prokaryotes

Control of gene expression in prokaryotes involves control of "operons".

Operon consists of:

1. promoter - DNA sequence where RNA polymerase (transcription enzyme) binds.
2. operator - DNA sequence where an active repressor protein binds.
3. structural genes - DNA coding for protein (usually several that are functionally related, e.g. enzymes of a metabolic pathway).

Regulatory components (NOT part of the operon):

1. regulatory gene - DNA sequence coding for production of a repressor protein (often located away from operon it regulates).
2. repressor protein - in active form, it binds to operator to prevent transcription of structural genes.

Repressible Operon - ex. tryptophan operon : Repressible operons are usually in the "on" condition, i.e. their gene products are generally required most of the time for cell maintenance, etc. Repressor protein is made in an inactive form (it needs to bind to a corepressor to make it an active repressor which can then bind to operator). Corepressor is usually an end-product of the metabolic pathway coded by the operon, e.g. tryptophan in our example here. Corepressor must accumulate to high levels before it can bind to inactive repressor and make it active.

Inducible Operon - ex. Lactose Operon

Inducible operons are usually in the "off" condition, i.e. their gene products are only occasionally (or never) needed by the cell. Repressor protein is made in an active form (capable of binding to operator by itself). Repressor protein can be inactivated by binding to an inducer (lactose in our example here) which then allows the operon to be activated.

Negative vs. Positive Controls

Above examples simply show operons being turned "on" or "off" at the operator, which represents a form of negative control. Some operons, when turned "on", do not get transcribed at a rate sufficient to see a noticable effect because the RNA polymerase does not bind to promoter effectively. These operons can be transcribed at a greatly accelerated rate by making the RNA polymerase binding much more efficient. This requires an activated "helper" for the RNA polymerase binding.

Gene Regulation in Eukaryotes

Developmental Biologists have long been interested in how a complex, multicellular organism is able to arise from a single fertilized egg (zygote), i.e. how cells arising from a common lineage (all arising from the zygote) are able to differentiate into the numerous, complex cell types that make up a multicellular organism. Since the zygote contains ALL the genes necessary for the development of an organism, how is it that only a limited number of these genes are expressed in certain cells so that a multitude of cell types arise from the zygote, e.g. expression of genes for nerve cells, bone cells, skin cells, etc. At one time, it was generally believed that progressive determination was caused by groups of cells actually losing most of the zygotes original set of genes until a particular cell type contained only genes for forming that cell type, e.g. a cell that eventually becomes a nerve cell has gradually lost all of the genes of the original zygote except those for nerve cell formation. Today we know that this is generally not the case, although there are exception.

Mitosis faithfully reproduces the full chromosome complement (and all of their genes) to be equally distributed to the new daughter cells, i.e. the daughter cells retain the total genetic potential of the original zygote, "totipotent cells". Much experimental evidence now supports this concept of the genetic equivalence of nuclei in a multicellular organism. Nuclear transplantation experiments performed by Robert Briggs and Thomas King, and John Gurdon showed that cells of early embryos retain full genetic potential of zygote. F.C. Steward was the first to show that individual cells taken from a mature carrot plant could grow into new carrot plants (clones).

More recently the cloning of "Dolly" the sheep, which was not all that important scientifically except that it showed that it could be done in mammals (humans?). Regeneration, in both plants and animals, have long supported the concept of totipotent cells or the genetic equivalence of somatic cell nuclei. Therefore, the fact that a differentiating cell is using only a limited number of the total available genes, the problem becomes a differential activation/inactivation of available genes.

For example, cells that eventually become nerve cells do so by activating only the genes necessary for nerve cell formation and likewise inactivating all other genes. The ways in which cells control gene expression has become a major focus of biological research in the last half of the 20th Century and with the more recent genetic engineering techniques will continue to be a major focus well into this century.

Most of what we first learned about differential gene activation/inactivation came from work on bacteria, e.g. *Escherichia coli*. Although these cells do not undergo the complex developmental phenomena associated with multicellular organism, they do require activation and inactivation of genes to maintain their existence, and their relative simplicity lent them very well to early studies of gene control. Many of the models for gene control in prokaryotes have been used as model systems to initially investigate eukaryotic gene control. However, as one might suspect, the controls used to regulate euakyotic genes appear to be much more complex and must take into account certain aspects of the eukaryote genome that are non-existent in prokaryotes.

1. Very large number of genes present (30,000+) human genes; although only about 3-5% is being used at any one time; recall that *E. coli* has only about 2500 genes.
2. Most eukaryote DNA does not code for RNA or protein (does not function as "genes").
3. Presence of chromatin and the structural complexity of the chromosome.
4. Compartmentalization of most DNA into a nucleus.
5. Complexity in variety of genes being activated at any one time (recall functionally linked genes into operons in bacteria), etc.

One way of looking at "gene expression" is to consider that a gene is expressed (or "activated") only when the product that the gene codes for is produced and functional, e.g. an "active" enzyme. Considering this very broad description for "gene activation", the eukaryotic cell possesses numerous places where control could take place. Some of these are, at present, only possibilities while others have good experimental support for their existence.

Potential Levels of Eukaryotic Gene Control

Pretranscriptional Control

The physical form of eukaryotic chromatin may be able to block the availability of genes for transcription by forming chromatin structures that "hide" the DNA from trancription factors and enzymes. "DNA packing" involved basic proteins called HISTONES (H1, H2A, H2B, H3, H4) and various other chromosomal proteins. An extreme form of DNA packing occurs when chromosomes condense during Prophase of cell division. Very little, if any, gene activity occurs while mitotic/meiotic chromosomes are formed. Heterochromatin is chromatin in a highly compacted form that is very low or completely devoid of transcriptional activity. It can occur in chromatin found in the Interphase cell. The tight coiling might be due to high methylation of the DNA in these areas. Regardless of the cause of heterochromatin, experimental evidence has shown very little transcriptional activity associciated with these areas. DNA methylation, in general, is a means of inactivating genes over extended time. An extreme example of gene inactivation by heterochromatin formation is the Barr body typically found in human females. In this example, an entire chromosome's genes are prevented from being transcribed (activated).

A chromosome may have heterochromatin areas interspersed with less compacted areas called euchromatin, which are thought to be areas where the DNA is more available for transcription. Even in euchromatin, the DNA is tightly wrapped around nucleosomes formed by histones. To make this DNA available for transcription, the histones of the nucleosome appear to undergo acetylation (addition of acetyl groups) which causes a "relaxation" of the DNA/ nucleosome association. This relaxing of the DNA/nucleosome assoication makes the DNA more available for transciption.

Transcriptional Control

Regulatory proteins have also been shown to exist in eukaryotes and may be either activators or repressors of gene activity. Since the majority of the eukaryotic genes are inactive at any one time, it is believed that most regulators are activators. Enhancers have also been shown to exist in eukaryotes. These are small stretches of DNA that have the ability to enhance transcriptional activity after binding activator proteins and then interacting at the promoter region to enhance RNA polymerase binding and activity. Recall that transcription factors are also needed for RNA polymerase binding and activity. Enhancers are found some distance away from the promoter and may be "upstream" or "downstream" from the promoter.

Posttranscriptional Control

The initial transcript (RNA) in eukaryotes is not functional but must undergo modification to become functional. RNA, tRNA, and mRNA must all undergo varying degrees of posttranscriptional modification before they can become functional.

In general, these modifications may include:

1. the cutting out of functional pieces of RNA from a much larger nonfunctional transcript.
2. some of the above pieces may be functional but some require the bonding together of excised pieces to form a single functional piece (exons/introns).
3. mRNA requires the most extensive posttranscriptional modifications, which in addition to the above, require chemical modifications at both ends of the molecule. These modifications may help the transport out of the nucleus, protect the molecule from nuclease attack in the cytoplasm, or somehow be required for initiation of translation.

mRNA is generally consider a "short-lived" molecule in both prokaryotes and eukaryotes, and was once used as a prime criterion for proving that a RNA was in fact a mRNA. However, exceptions to this generality have been shown in eukaryotes, particularly in certain highly specialized cells (e.g. maturing vertebrate erythrocytes) and developing embryos. In the case of the maturing vertebrate red blood cell (called erythropoiesis), the erythroblast is a nucleated cell that gradually shuts down most of it's genes except for mRNA for globin production. Huge amounts of globin mRNA accumulate and eventually the cell becomes anucleate, however, the globin mRNA remains for a considerable time afterwards while the cell makes and accumulates very large amounts of hemoglobin. Unfertilized eggs have been shown to manufacture and accumulate large amounts of mRNA coding for proteins necessary for the rapid, early development of the embryo; these mRNAs are stored in an inactive form in the egg. Activation of the egg, by sperm or artificial means, activates these mRNAs for embryo protein production.

Translational Control

The observation that eggs may contain "long-lived" mRNA but that this RNA is not translated until egg activation has been shown to be a block at translation. A certain initiation factor(s) necessary for translation rapidly becomes available after fertilization (activation). Also, the amount of heme remaining in maturing red blood cells helps determine how much more globin will be made (translation of globin mRNA). The poliovirus is thought to select the translation of its "uncapped" mRNAs by somehow destroying the host cell's cap recognition system on the ribosome complex.

Posttranslational Control

Even after the protein is first made, it is not necessarily functional. Insulin is initially made as a single protein (inactive proinsulin) which must fold back upon itself, form three disulfide bonds, and finally have a small stretch of amino acids removed by a peptidase before it becomes active insulin consisting of two polypeptides held together by two of the disulfide bonds. Many of the digestive enzymes of the small intestine are synthesized as zymogens (inactive forms) that must undergo enzymatic cleavage to become active, e.g. trypsinogen/trypsin, chymotrypsinogen/chymotrypsin, pepsinogen/pepsin. Some proteins (malformed, no longer needed, etc.) are "marked" for degradation by proteosomes.

Eukaryotic Gene Organization and its Evolution

Unlike the prokaryotes, eukaryotes do not possesss an "operon type" arrangement of functionally related genes. (recall prokaryote operon). Instead, it has been shown that genes coding for functionally related products, e.g. a series of enzymes in a metabolic pathway, are scattered throughout the genome and each controlled by a separate promoter, etc. This gene arrangement may at first appear disorganized but the reality is that it allows much greater variety in the use of these "scattered" genes. Genes can be activated in numerous combinations. A multigene family is a collection of genes within a genome that are similar or identical in sequence. The 28S, 18S and 5.8S rRNA genes occur in eukaryotes as tandem arrays of several hundred to several thousand located in association with a nucleosome found inside the nucleus.

The family of globin genes are sequence related but not identical and function in making different hemoglobins at different developmental stages. Both types of multigene families most likely arose by gene duplication and mutation.

Molecular Biology of Cancer

Cancer is caused by inapproriate gene activity associated with the cell cycle.

Some chararcteristics of malignant cancers (solid types):

1. Loss of cell cycle control
2. Cell surface alterations
3. Secretion of distinct products, e.g. proteolytic enzymes, embryonic antigens, angiogenin, etc.
4. Morphological alterations, enlarged nuclei, aberrant nucleoli, numerous microvilli
5. Highly vascularized
6. Metastasis (travel to distant tissues and establish secondary or "metastatic" cancers).

The genetic basis of cancer was first discovered through the work with retroviruses that were shown to cause cancer in monkeys, chickens, mice, etc. The viral genes were identified that could be directly linked to the cause of cancer in experimental animals. These genes were called oncogenes ("cancer genes"). Upon closer examination, researchers found that in fact the oncogenes were not unique but were in fact mutated forms of normal genes (proto-oncogene) found in eukaryotic organisms; including humans. These normal genes, for the most part, are involved with some aspect of regulating the cell cycle.

Three ways for conversion of a proto-oncogene into an oncogene:

1. translocation or transposition to new promoter
2. gene amplification
3. point mutation within gene.

In addition to proto-oncogenes that can become oncogenes, there are normal, tumour-suppressor genes that are mutated during the formation of cancers. Tumour-suppressor genes produce products that normally block the cell cycle; which would normally prevent cancers. The cancerous state results from alteration of several genes, both proto-oncogenes and tumour suppressor genes.

7

Gene Expression

The cloning of Dolly, a sheep, was reported worldwide in 1996. Dolly developed from adult somatic nuclei that had been implanted into enucleated eggs (eggs with the nuclei removed). More recently, cows, pigs, mice, and other mammals have been cloned as well with the use of similar technology. The successful cloning of Dolly was a great surprise to the scientific community because the cloning of mammals from somatic cells was thought to be impossible. A reason for the initial skepticism was that the formation of male and female gametes (sperm and egg cells) was known to include sex-specific modifications to the respective genomes that resulted in sex-specific patterns of gene expression. As such, Dolly is symbolic of how far we have progressed in understanding aspects of eukaryotic gene regulation such as the global control of gene expression exemplified by gamete development. However, for every successful clone, including Dolly, there are many more, perhaps hundreds of embryos that fail to develop into viable progeny. The extremely high failure rate underscores how much remains to be deciphered about eukaryotic gene regulation.

The biological properties of each eukaryotic cell type are largely determined by the proteins expressed within it. This constellation of expressed proteins determines much of the cell's architecture, its enzymatic activities, its interactions with its environment, and many other physiological properties. However, at any given time in a cell's life history, only a fraction of the RNAs and proteins encoded in its genome are expressed. At different times, the profile of expressed gene products can differ dramatically, both in regard to which proteins are expressed and at what levels. How are these specific profiles generated? As one might expect, if the final product is a protein, regulation could be achieved by controlling the transcription of DNA into RNA or the translation of RNA into protein.

In fact, gene regulation takes place at many levels, including at the mRNA level (through alterations in splicing or the stability of the mRNA) and after translation (by modifications of proteins). However, most regulation is thought to take place at the level of gene transcription; so, in this chapter, the primary focus is on the regulation of transcription. The basic mechanism at work is that molecular signals from outside or inside the cell lead to the binding of regulatory proteins to specific DNA sites outside of protein-encoding regions, and the binding of these proteins modulates the rate of transcription. These proteins may directly or indirectly assist RNA polymerase in binding to its transcription initiation site—the *promoter*—or they may repress transcription by preventing the binding of RNA polymerase.

Although bacteria and eukaryotes have much of the logic of gene regulation in common, there are some fundamental differences in the underlying mechanisms and machinery. Both use sequence-specific DNA-binding proteins to modulate the level of transcription. However, eukaryotic genomes are bigger and their range of properties is larger than those of bacteria. Inevitably their regulation is more complex, requiring more types of regulatory proteins and more types of interactionswith the adjacent regulatory regions in DNA. The most important difference is that eukaryotic DNA is packaged into *nucleosomes,* forming *chromatin,* whereas bacterial DNA lacks nucleosomes. In eukaryotes, chromatin structure is dynamic and is an essential ingredient in gene regulation.

In general, the ground state of a bacterial gene is "on." Thus, RNA polymerase can usually bind to a promoter when no other regulatory proteins are around to bind to the DNA. In bacteria, transcription initiation is prevented or reduced if the binding of RNA polymerase is blocked, usually through the binding of a repressor regulatory protein. Activator regulatory proteins increase the binding of RNA polymerase to promoters where a little help is needed. In contrast, the ground state in eukaryotes is "off." Therefore, the transcriptional machinery (including RNA polymerase II and associated general transcription factors) cannot bind to the promoter in the absence of other regulatory proteins. In many cases, the binding of the transcriptional apparatus is not possible, owing to the position of nucleosomes near the promoter. Thus, chromatin structure usually has to be changed to activate eukaryotic transcription. The structure of chromatin around activated or repressed genes within cells can be quite stable and inherited by daughter cells. The inheritance of chromatin structure is a form of inheritance that does not directly entail DNA sequence.

1. In bacteria, all genes are transcribed into RNA by the same RNA polymerase, whereas three RNA polymerases function in eukaryotes.

2. RNA transcripts are extensively processed during transcription in eukaryotes; the 5_ and 3_ ends are modified and introns are spliced out.
3. RNA polymerase II is much larger and more complex than its bacterial counterpart. One reason for the added complexity is that RNA polymerase II must synthesize RNA *and* coordinate the special processing events unique to eukaryotes.

Multicellular eukaryotes may have as many as 25,000 genes, severalfold more than the average bacterium. Moreover, patterns of eukaryotic gene expression can be extraordinarily complex. That is, there is great variation among genes in when a gene is on (transcribed) or off (not transcribed) and in how much transcript needs to be made. For example, one gene may be transcribed only during early development and another only in the presence of a viral infection. Finally, the majority of the genes in a eukaryotic cell are off at any one time. On the basis of these considerations alone, eukaryotic gene regulation must be able to ensure that the expression of most genes in the genome is off at any one time while activating a subset of genes; and generate thousands of patterns of gene expression.

Mechanisms have evolved to ensure that most of the genes in a eukaryotic cell are not transcribed. Before considering how genes are kept transcriptionally inactive, we will focus on the second point: How are eukaryotic genes able to exhibit an enormous number and diversity of expression patterns? The machinery required for generating so many patterns of gene transcription in vivo has many components, including both regulatory proteins and cis-acting regulatory sequences. To initiate transcription, these proteins interact with DNA sequences called promoter-proximal elements near the promoter of a gene. The second group of protein components consists of specific transcription factors that bind to cis-acting regulatory sequences in the DNA called enhancers or upstream activating sequences (UAS's). These regulatory sequences may be located a considerable distance from gene promoters. Generally speaking, promoters and promoter-proximal elements are bound by transcription factors that affect the expression of many genes. Enhancers are the targets of more specific transcription factors that control the regulation of smaller subsets of genes. Often, an enhancer will act in only one or a few cell types in a multicellular eukaryote.

For RNA polymerase II to transcribe DNA into RNA at a maximum rate, multiple cis-acting regulatory elements must play a part. The promoters, promoter-proximal elements,and enhancers are all targets for binding by different trans-acting DNA binding proteins. The binding of RNA polymerase II to the promoter does not produce efficient transcription by itself. Transcription requires the binding of general transcription factors to additional promoter-proximal elements that are commonly found within 100 bp of the transcription initiation site of many (but not all) genes. One of these elements is the CCAAT box, and often another is a GC-rich segment farther upstream.

The general transcription factors that bind to the promoter-proximal elements are expressed in most cells, and sothey are available to initiate transcription at any time. Mutations in these sites can have a dramatic effect on transcription, demonstrating how important they are.

To modulate transcription, regulatory proteins possess one or more of the following functional domains:

1. A domain that recognizes a DNA regulatory sequence (the protein's DNA-binding site)
2. A domain that interacts with one or more proteins of the transcriptional apparatus (RNA polymerase or a protein associated with RNA polymerase)
3. A domain that interacts with proteins bound to nearby regulatory sequences on DNA such that they can act cooperatively to regulate transcription
4. A domain that influences chromatin condensation either directly or indirectly
5. A domain that acts as a sensor of physiological conditions within the cell Much of the strategy of eukaryotic transcriptional control hinges on how specific transcription factors control the access of general transcription factors and RNA polymerase II. Eukaryotic gene regulatory mechanisms have been discovered through both biochemical and genetic approaches. The latter has been advanced in particular by studies of the single-celled yeast *Saccharomyces cerevisiae*. Several decades of research have been a source of many insights into general principles of how eukaryotic transcriptional regulatory proteins work and how different cell types are generated. We'll examine two yeast gene regulatory systems in detail: the first concerns the galactose-utilization pathway; the second is the control of mating type.

Model Organism *Yeast*

Saccharomyces cerevisiae, or budding yeast, has emerged in recent years as the premier eukaryotic genetic system. Humans have grown yeast for centuries because it is an essential component of beer, bread, and wine. Yeast has many features that make it an ideal model organism. As a unicellular eukaryote, it can be grown on agar plates and, with a life cycle of just 90 minutes, large quantities can be cultured in liquid media. It has a very compact genome with only about 12 megabase pairs of DNA (compared with almost

3000 megabase pairs for humans) containing approximately 6000 genes that are distributed on 16 chromosomes. It was the first eukaryote to have its genome sequenced.

The yeast life cycle makes it very versatile for laboratory studies. Cells can be grown as either diploid or haploid. In both cases, the mother cell produces a bud containing an identical daughter cell. Diploid cells either continue to grow by budding or are induced to undergo meiosis, which produces four haploid spores held together in an *ascus* (also called a *tetrad*). Haploid spores of opposite mating type (a or ‘) will fuse and form a diploid. Spores of the same mating type will continue growth by budding. Yeast has been called the *E. coli* of eukaryotes because of the ease of forward and reverse mutant analysis. To isolate mutants by using a forward genetic approach, haploid cells are mutagenized (with X rays, for example) and screened on plates for mutant phenotypes. This procedure is usually done by first plating cells on a rich medium on which all cells grow and by copying, or *replica plating,* the colonies from this master plate onto replica plates containing selective media or special growth conditions. For example, temperature-sensitive mutants will grow on the master plate at the permissive temperature but not on a replica plate at a restrictive temperature. Comparison of the colonies on the master and replica plates will reveal the temperature-sensitive mutants. Using reverse genetics, scientists can also replace any yeast gene (of known or unknown function) with a mutant version (synthesized in a test tube) to understand the nature of the gene product.

Lessons from Yeast: The GAL System

To make use of extracellular galactose, yeast imports the sugar and converts it into a form of glucose that can be metabolized. Several genes—*GAL1, GAL2, GAL7,* and *GAL10*—in the yeast genome encode enzymes that catalyze steps in the biochemical pathway that converts galactose into glucose. Three additional genes—*GAL3, GAL4,* and *GAL80*—encode proteins that regulate the expression of the enzyme genes. Just as in the *lac* system, the abundance of the sugar determines the level of gene expression in the biochemical pathway. In yeast cells growing in media lacking galactose, the *GAL* genes are largely silent. But, in the presence of galactose (and the absence of glucose), the *GAL* genes are induced. Just as for the *lac* operon, genetic and molecular analyses of mutants have been key to understanding how the expression of the genes in the galactose pathway is controlled. The key regulator of *GAL* gene expression is the Gal4 protein, a sequencespecific DNA-binding protein. Gal4 is perhaps the best-studied transcriptional regulatory protein in eukaryotes. The detailed dissection of its regulation and activity has been a source of several key insights into the control of transcription in eukaryotes.

Gal4 Regulates Multiple Genes through Upstream Activating Sequences

In the presence of galactose, the *GAL1, GAL2, GAL7,* and *GAL10* genes are induced 1000-fold or more. In *GAL4* mutants, however, they remain silent. Each of these four genes has two or more Gal4-binding sites located 5_ (upstream) of its promoter. Consider the *GAL10* and *GAL1* genes, which are adjacent to each other and transcribed in opposite directions. Between the *GAL1* transcription start site and the *GAL10* transcription start site is a single 118-bp region that contains four Gal4- binding sites. Each Gal4-binding site is 17 base pairs long and is bound by one Gal4 protein dimer. There are two Gal4-binding sites upstream of the *GAL2* gene as well, and another two upstream of the *GAL7* gene. These binding sites are required for gene activation in vivo. If they are deleted, the genes are silent, even in the presence of galactose. These regulatory sequences are enhancers that are also referred to as upstream activating sequences. The presence of enhancers located at a considerable linear distance from a eukaryotic gene's promoter is typical.

The level of gene expression in the biochemical pathway. In yeast cells growing in media lacking galactose, the *GAL* genes are largely silent. But, in the presence of galactose (and the absence of glucose), the *GAL* genes are induced. Just as for the *lac* operon, genetic and molecular analyses of mutants have been key to understanding how the expression of the genes in the galactose pathway is controlled. The key regulator of *GAL* gene expression is the Gal4 protein, a sequencespecific DNA-binding protein. Gal4 is perhaps the best-studied transcriptional regulatory protein in eukaryotes. The detailed dissection of its regulation and activity has been a source of several key insights into the control of transcription in eukaryotes.

The Gal4 protein has separable DNA-binding and Activation Domains

After Gal4 is bound to the UAS element, how is gene expression induced? A distinct domain of the Gal4 protein, the activation domain, is required for regulatory activity. Thus, the Gal4 protein has at least two domains: one for DNA binding and another for activating transcription. A similar modular organization has been found to be a common feature of other DNA-binding transcription factors as well. The modular organization of the Gal4 protein was demonstrated in a series of simple, elegant experiments. The strategy was to test the DNA binding and gene activation of mutant forms of the protein in which parts had been either deleted or fused to other proteins. In this manner, whether a part of the protein was necessary for a particular function could be determined. To carry out these studies, experimenters needed a simple means of assaying the expression of the enzymes encoded by the *GAL* genes. The expression of *GAL* genes and other targets of transcription factors is typically monitored by using a reporter gene whose expression is easily tracked. Often, the reporter gene is the *lacZ* gene of *E. coli,* which can act on substrates whose products are easily measured by their bright color or fluourescence. Another common reporter gene is the gene that encodes the green fluorescent protein of jellyfish, which, as its name suggests, is easily tracked by the light that it emits. The coding region of one of these reporter genes and a promoter are placed downstream of a UAS element from a *GAL* gene.

Reporter expression is then a read-out of Gal4 activity in cells. When a form of the Gal4 protein lacking the activation domain is expressed in yeast, the binding sites of the UAS element are occupied, but no transcription is stimulated. The same is true when other regulatory proteins lacking activation domains, such as the bacterial repressor LexA, are expressed in cells bearing reporter genes with their respective binding sites. The more interesting result is obtained when a form of the Gal4 protein lacking the DNA-binding domain is grafted to the LexA DNA-binding domain; the hybrid protein now activates transcription from LexAbinding sites. Further "domain swap" experiments have revealed that the transcriptional activation function of the Gal4 protein resides in two small domains about 50 to 100 amino acids in length. These domains are separable from those used in the dimerization of the protein, DNA binding, and interaction with the Gal80 protein. The activation domain helps recruit the transcriptional machinery to the promoter. This highly modular arrangement of activityregulating domains is found in many transcription factors.

Gal4 Activity is Physiologically Regulated

How does Gal4 become active in the presence of galactose? Key clues came from analyses of mutations in the *GAL80* and *GAL3* genes. In *GAL80* mutants, the *GAL* structural genes are active even in the absence of galactose. This result suggests that the normal function of the Gal80 protein is to somehow inhibit *GAL* gene expression. Conversely, in *GAL3* mutants, the *GAL* structural genes are not active in the presence of galactose, suggesting that Gal3 normally promotes expression of the *GAL* genes.

Extensive biochemical analyses have revealed that the Gal80 protein binds to the Gal4 protein with high affinity and directly inhibits Gal4 activity. Specifically, Gal80 binds to a region within one of the Gal4 activation domains, blocking its ability to promote the transcription of target genes. The role of the Gal3 protein is to release Gal4 from its inhibition by Gal80 in the presence of galactose. Gal3 is a sensor and inducer. When Gal3 binds galactose and ATP, it undergoes an allosteric change that promotes binding to Gal80, which in turn causes Gal80 to release Gal4, which is then able to activate transcription of its target genes. Thus, Gal3, Gal80, and Gal4 are all part of a switch whose state is determined by the presence or absence of galactose. In this switch, DNA binding by the transcriptional regulator is not the physiologically regulated step (as is the case in the *lac* operon and bacteriophage); rather, the activity of the activation domain is regulated.

Gal4 Functions in most Eukaryotes

In addition to its action in yeast cells, Gal4 has been shown to be able to activate transcription in insect cells, human cells, and many other eukaryotic species. This versatility suggests that biochemical machinery and mechanisms of gene activation are common to a broad array of eukaryotes and that features revealed in yeast are generally present in other eukaryotes, and vice versa. Furthermore, because of their versatility, Gal4 and its UAS elements have become favored tools in genetic analysis for manipulating gene expression and function in a wide variety of model systems.

Activators Recruit the Transcriptional Machinery

In bacteria, activators commonly stimulate transcription by interacting directly with DNA and with RNA polymerase. In eukaryotes, activators generally work indirectly to recruit RNA polymerase II to gene promoters through two major mechanisms. First, activators can interact with subunits of the protein complexes having roles in transcription initiation. Second, activators can recruit proteins that modify chromatin structure, allowing RNA polymerase II and other proteins access to the DNA. Many activators, including Gal4, have both activities. The eukaryotic transcriptional machinery contains many proteins that are parts of various subcomplexes within the transcriptional apparatus that is assembled on gene promoters. One subcomplex, transcription factor IID (TFIID), binds to the TATA box of eukaryotic promoters through the TATAbinding protein. Gal4 binds to TBP at a site in its activation domain, and, through this binding, it recruits the TFIID complex and, in turn, RNA polymerase II to the promoter. The affinity of this interaction correlates well with Gal4's potency as an activator. Gal4 also interacts with the large Mediator complex, which directly interacts with RNA polymerase II to recruit it to gene promoters. The Mediator complex is an example of a coactivator, a term applied to a protein or protein complex that facilitates gene activation by a transcription factor but that itself is neither part of the transcriptional machinery nor a DNA-binding protein. The ability of activators to bind to upstream DNA sequences and to interact with proteins that bind directly or indirectly to promoters helps to explain how transcription can be stimulated from more distant regulatory sequences.

Dynamic Chromatin and Eukaryotic Gene Regulation

A second mechanism for influencing gene transcription in eukaryotes modifies the local chromatin structure around gene regulatory sequences. To fully understand how this mechanism works, we need to first review chromatin structure and then consider how it can change and how these changes affect gene expression. The recruitment of transcriptional machinery by activators may appear to be somewhat similar in eukaryotes and bacteria, with the major difference being in the number of interacting proteins in the transcriptional machinery. Indeed, less than a decade ago, many biologists pictured eukaryotic regulation simply as a biochemically more complicated version of what had been discovered in bacteria. However, this view has changed dramatically as biologists have considered the effect of the organization of genomic DNA in eukaryotes.

Compared with eukaryotic DNA, bacterial DNA is relatively "naked," making it readily accessible to RNA polymerase. In contrast, eukaryotic chromosomes are packaged into chromatin, which is composed of DNA and proteins (mostly histones). The basic unit of chromatin is the nucleosome, containing about 150 bp of DNA wrapped twice around a histone octamer. The histone octamer is composed of two subunits of each of the four histones: histone 2A, 2B, 3, and 4. Nucleosomes can associate into higherorder structures that further condense the DNA. The packaging of eukaryotic DNA into chromatin means that much of the DNA is not readily accessible to regulatory proteins and the transcriptional apparatus. Thus, whereas prokaryotic genes are generally accessible and "on" unless repressed, eukaryotic genes are inaccessible and "off" unless activated. Therefore, the modification of chromatin structure is a distinctive feature of eukaryotic gene regulation.

One can imagine several ways to alter chromatin structure. For example, one mechanism might be to simply move the histone octamer along the DNA. In the 1980s, biochemical techniques were developed that allowed researchers to determine the position of nucleosomes in and around specific genes. In these studies, chromatin was isolated from tissues or cells in which a gene was on and compared with chromatin from tissue where the same gene was off. The result for most genes analyzed was that nucleosome positions changed, especially in a gene's regulatory regions. Thus, which DNA regions are wrapped up in nucleosomes can change: Nucleosome positions can shift on the DNA from cell to cell and over the life cycle of an organism.

Transcription might be repressed when the promoter and flanking sequences are wound up in a nucleosome and inaccessible to RNA polymerase II. Activation of transcription would thus require the blocking nucleosome to be reorganized by nudging the histones or removing them entirely. Conversely, when gene repression is necessary, histone octamers may shift into a position that prevents transcription. The changing of nucleosome position is referred to as chromatin remodeling. Now, chromatin remodeling is known to be an integral part of eukaryotic gene expression, and great advances are being made in determining the underlying mechanism(s) and the regulatory proteins taking part. Here, again, genetic studies in yeast have been pivotal.

Chromatin-remodeling Proteins and Gene Activation

What was the connection between the ability to utilize sugar and the ability to switch mating types? The Snf2–Swi2 protein was purified and discovered to be part of a large, multisubunit complex called the SWI–SNF complex that can reposition nucleosomes in a test-tube assay if ATP is provided as an energy source. In some situations, the multisubunit SWI–SNF complex activates transcription by moving nucleosomes that are covering the TATA sequences and, in this way, facilitates the binding of RNA polymerase II. The SWI–SNF complex is thus a coactivator.

Gal4 also binds to the SWI–SNF complex and recruits the chromatin-remodeling complex to activated promoters. Yeast strains containing a defective SWI–SNF complex show a reduced level of Gal4 activity. Why might an activator use multiple activation mechanisms? There are at least two reasons understood at present. The first is that the accessibility of target promoters may change at different stages of the cell cycle or in different cell types (in multicellular eukaryotes). For example, during mitosis, when chromatin is more condensed, genes are less accessible. At that stage, Gal4 must recruit the chromatin-remodeling complexes, whereas, at other times, such recruitment might not be required to activate gene expression. A second reason is that many transcription factors act in combinations to control gene expression synergistically. We will see shortly that this combinatorial synergy is a result of the fact that chromatin-remodeling complexes and the transcriptional machinery are recruited more efficiently when multiple transcription factors act together.

Histones and Chromatin Remodeling

A histone code As already stated, most nucleosomes are composed of an octamer made up of two copies each of the four core histones. Histones are known to be the most conserved proteins in nature; that is, histones are almost identical in all eukaryotic organisms from yeast to plants to animals. This conservation contributed to the view that histones could not take part in anything more complicated than the packaging of DNA to fit in the nucleus. However, recall that DNA with its four bases also was considered too "dumb" a molecule to carry the blueprint for all organisms on Earth. Of note is that the histone proteins are organized into the core octamer with their amino-terminal ends protruding from the nucleosome. These protruding ends are called histone tails. Since the early 1960s, specific lysine residues in the histone tails have been known to be able to be covalently modified by the attachment of acetyl and methyl groups. These reactions take place after the histone protein has been translated and even after the histone has been incorporated into a nucleosome. There are now known to be at least 150 different histone modifications that require a wide variety of molecules in addition to the acetyl and methyl groups already mentioned (for example, phosphorylation and ubiquitylation). Histone acetylation, deacetylation, and gene expression The acetylation reaction is the best-characterized histone modification:

Note that the reaction is reversible, which means that acetyl groups can be added and removed from the same histone residue. With 44 histone lysine residues available to accept acetyl groups, the presence or absence of these groups can carry a tremendous amount of information. For this reason, the covalent modification of histone tails is said to be a histone code. Scientists coined the expression histone code because the covalent modification of histone tails is reminiscent of the genetic code. For the histone code, information is stored in the patterns of histone modification rather than in the sequence of nucleotides. With more than 150 known histone modifications, there are a huge number of possible patterns and their effects on chromatin structure and transcriptional regulation are just beginning to be deciphered. To add to this complexity, the code is likely not interpreted in precisely the same way in all organisms. For now, let's see how the acetylation of histone amino acids influences chromatin structure and gene expression.Evidence had been accumulating for years that the histones associated with the nucleosomes of active genes are rich in acetyl groups (said to be hyperacetylated), whereas inactive genes are underacetylated (hypoacetylated). The enzyme responsible for adding acetyl groups, histone acetyltransferase (HAT), proved very difficult to isolate. When it was finally isolated and its protein sequence deduced, it was found to be an *ortholog* of a yeast transcriptional activator called GCN5 (meaning that it was encoded by the same gene in a different organism).

Thus, the conclusion was that GCN5 is a histone acetyltransferase. It binds to the DNA in the regulatory regions of some genes and activates transcription by acetylating nearby histones. Various protein complexes that are recruited by transcriptional activators are now understood to possess a HAT activity. How does histone acetylation facilitate changes in gene expression? There appear to be at least two mechanisms for doing so. First, the addition of acetyl groups to specific histone residues can alter the interaction in a nucleosome between the DNA and a histone octamer so that the octamer is more likely to slide along the DNA to a new position. Second, histone acetylation, in conjunction with other histone modifications, influences the binding of regulatory proteins to the DNA. The bound regulatory protein may take part in one of several functions that either directly or indirectly increase the frequency of transcription initiation. Like other histone modifications, acetylation is reversible, and histone deacetylases (HDAT's) also have been identified. Such proteins play key roles in gene repression. For example, in the presence of galactose and glucose, the activation of *GAL* genes is prevented by the Mig1 protein. Mig1 is a sequence-specific DNA-binding repressor that binds to a site between the UAS element and the promoter of the *GAL1* gene. Mig1 recruits a protein complex called Tup1 that contains a histone deacetylase and that represses gene transcription. The Tup1 complex is an example of a corepressor, which faciliates gene repression but is not itself a DNA-binding repressor. The Tup1 complex is also recruited by other yeast repressors, such as MAT?2, and counterparts of this complex are found in all eukaryotes.

Mechanism of Enhancer Action

The development of a complex organism requires that transcription levels be regulated over a wide range. Think of a regulation mechanism as more like a rheostat than an on-or-off switch. In eukaryotes, transcription levels are made finely adjustable by the clustering of binding sites into enhancers. Several different transcription factors or several molecules of the same transcription factor may bind to adjacent sites. The binding of these factors to sites that are the correct distance apart leads to an amplified, or superadditive, effect on activating transcription. When an effect is greater than additive, it is said to be synergistic. The binding of multiple regulatory proteins to the multiple binding sites in an enhancer can catalyze the formation of an enhanceosome, a large protein complex that acts synergistically to activate transcription. To better understand what an enhanceosome is and how it acts synergistically, let's look at a specific example.

The a-interferon Enhanceosome

The human -interferon gene, which encodes the antiviral protein interferon, is one of the best-characterized genes in eukaryotes. It is normally switched off but is activated to very high levels of transcription on viral infection. The key to the activation of this gene is the assembly of transcription factors into an enhanceosome about 100 bp upstream of the TATA box and transcription start site. The regulatory proteins of the -interferon enhanceosome all bind to the same face of the DNA double helix. Binding to the other side of the helix are several architectural proteins that bend the DNA and allow the different regulatory proteins to touch one another and form an activated complex. When all of the regulatory proteins are bound and interacting correctly, they form a "landing pad," a highaffinity binding site for the protein CBP, a coactivator protein that also recruits the transcriptional machinery. The large CBP protein also contains an intrinsic histone acetylase activity that modifies nucleosomes and facilitates high levels of transcription. Cooperative interactions help to explain several perplexing observations about enhancers. For example, they explain why mutating any one transcription factor or binding site dramatically reduces enhancer activity. They also explain why the distance between binding sites within the enhancer is such a critical feature. Furthermore, enhancers do not have to be close to the start site of transcription.

The Control of Yeast Mating Type: Combinatorial Interactions

Thus far, we have focused in this chapter on the regulation of single genes or a few genes in one pathway. In multicellular organisms, distinct cell types differ in the expression of hundreds of genes. The expression or repression of sets of genes must therefore be coordinated in the making of particular cell types. One of the best-understood examples of cell-type regulation in eukaryotes is the regulation of mating type in yeast. This regulatory system has been dissected by an elegant combination of genetics, molecular biology, and biochemistry. Mating type serves as an excellent model for understanding the logic of gene regulation in multicellular animals. The yeast *Saccharomyces cerevisiae* can exist in any of three different cell types known as a, and 'a/'. The two cell types a and ' are haploid and contain only one copy of each chromosome. Although the two haploid cell types cannot be distinguished by their appearance in the microscope, they can be differentiated by a number of specific cellular characteristics, principally their mating type. An cell mates only with an a cell and secretes an oligopeptide pheromone, or sex hormone, called factor that arrests a cells in the cell cycle. A cell of the a type mates only with an cell and secretes a pheromone, called a factor, that arrests cells. The diploid a cell does not mate, is larger than the and a cells, and does not respond to the mating hormones. Genetic analysis of mutants defective in mating has shown that cell type is controlled by a single genetic locus, the mating-type locus, *MAT*. There are two alleles of the *MAT* locus: haploid a cells have the *MATa* allele, haploid 'cells have the *MATa* allele, and the a/' diploid has both alleles. Although mating type is under genetic control, certain strains switch their mating type, sometimes as frequently as every cell division.

Enhancer-blocking Insulators

A regulatory element, such as an enhancer, that can act over tens of thousands of base pairs could interfere with the regulation of nearby genes. To prevent such promiscuous activation, regulatory elements called enhancer-blocking insulators have evolved. When positioned between an enhancer and a promoter, enhancerblocking insulators prevent the enhancer from activating transcription at that promoter. Such insulators have no effect on the activation of other promoters that are not separated from their enhancers by the insulator. Several models have been proposed to explain how an insulator could block enhancer activity only when placed between an enhancer and a promoter.

8

Gene Development

Health-care professionals who specialize in genetic disorders use a variety of methods to identify inherited conditions. Analysis of a family medical history, known as pedigree analysis, is used to track the transmission of a condition through generations. Blood tests that identify specific DNA sequences can reveal carriers of a disease-causing gene who have no symptoms of the disease. Geneticists collect a person's medical family history to trace the inheritance of a genetic trait among multiple generations.

The information is placed in a pedigree, which resembles a traditional multigenerational family tree but includes information about individuals who were diagnosed with a particular disorder or who suffered from certain medical symptoms. A pedigree can help researchers recognize diseases that express themselves in dominant or recessive alleles. Dominant disorders affect every generation. Recessive disorders may cluster in a single generation, reflecting when two parents who both carry a recessive allele for a disease have one or more children who develop the disease. A pedigree can also identify diseases that show Y-linked inheritance. Pedigree analysis can be useful when combined with certain genetic tests. A blood sample taken from a person who is at risk for a genetic disorder can be compared with a DNA sequence known to cause the disorder in question. Other genetic tests can reveal if a person has extra chromosomes, missing chromosomes, or chromosomes that have attached to one another in unusual ways. In some cases, these chromosomal abnormalities may produce genetic disorders in children or they may affect a person's ability to conceive a child. Genetic testing can also identify disorders in a fetus, enabling parents to learn early in a pregnancy if a fetus will likely be born with health problems or develop them later in life. Presymptomatic testing can identify DNA abnormalities in a person before health problems develop. In the case of certain inherited heart conditions, for example, these tests enable a person to make healthy lifestyle changes or take other preventative measures, such as medications, to lower the risk of illness or death.

Medical genetic testing raises challenging issues because such tests typically provide statistical possibilities rather than a definite prediction of whether a person will develop a given genetic disease. A test result may indicate, for example, that a person has a 75 per cent risk of developing colon cancer by the age of 65. Such results enable a physician to perform appropriate screening tests on the at-risk person in order to identify the disease at its earliest stages, when it is most treatable. At the same time, however, physicians must decide at what age the person's screening should begin and whether the benefits of early screening are worth the drawbacks of frequent screening. These drawbacks include expense, patient anxiety and discomfort, and exposure to radioactivity or other harmful substances used in testing.

Different problems are posed by genetic screening tests that diagnose conditions for which no preventive measures exist, such as Alzheimer's disease, a progressive brain disorder that causes the loss of mental function. People may find it devastating to learn that they are at risk for a deadly disease that cannot be prevented by medical measures or lifestyle choices. A recent development in genetic technology known as gene therapy focuses on curing inherited disorders. In experiments using gene therapy, researchers have replaced defective genes with normal alleles, inactivated a mutated gene, or inserted a normal form of a gene into a chromosome. The earliest success in human gene therapy involved the treatment of infants who cannot produce adenosine deaminase (ADA), an enzyme important to normal function of the immune system. Scientists have successfully inserted the normal allele for the gene that codes for the enzyme into cells in ADA-deficient children. Preliminary evidence indicates that this gene therapy leads to better immune function in recipients. Researchers are also exploring gene therapy's potential to help treat people with many other conditions, including certain cancers, hemophilia, heart disease, and cystic fibrosis. Although the United States Food and Drug Administration (FDA) has approved more than 400 clinical trials in gene therapy, this method of treating disease remains far from an unqualified medical success.

Treatments usually produce some improvement in the underlying condition, but not enough to consider the therapy suitable for large-scale use. The death of a patient involved in a gene therapy experiment in 1999 caused the National Institutes of Health (NIH), a federal agency that monitors gene therapy studies, to reevaluate the safety and effectiveness of gene therapy clinical trials.

Human Genome Project

The Human Genome Project is the most ambitious project in the history of biology. The programme's challenging goal was to identify and sequence all of the DNA in human chromosomes. The project was initiated in 1990 in the United States with government funding, and it rapidly grew into an international consortium of academic centres and drug companies in China, France, Germany, Japan, the United Kingdom, and the United States.

The consortium initially hoped to reach its goal by the year 2005. In 1998 Celèra Genomics, a privately funded biotechnology firm, announced that it would sequence the human genome by the year 2000 using different sequencing strategies than those used by the public consortium. This announcement triggered a heated race between Celera Genomics and the public consortium to complete the genome project. In June 2000 both teams declared victory when they jointly announced that they had separately completed a rough draft of the genome.

The two teams published their findings simultaneously, although in two different journals, in February 2001. The draft provided a basic outline of 90 per cent of the human genome. Scientists from the public consortium completed the final sequencing of the human genome in April 2003, two years earlier than planned. The completed human genome has provided scientists with a detailed blueprint of our complex genetic code. Large computer databases of genetic information enable scientists to look for patterns and relationships among the actions of different genes.

Among the findings about the human genome was that the number of genes in the human genome is much lower than was predicted—only about 20,000 to 25,000 genes compared to the expected 100,000 genes. This number is a little more than twice the number of genes found in the fruit fly. Scientists are now turning their attention to studying how the relatively low number of genes in the human genome can produce the complex structures found in humans.

Scientists have long known that a single gene produces a single protein and that this single protein subsequently may be processed into several different proteins. In a new science known as proteomics, scientists seek to identify and understand the function of all the proteins in the human body. They theorize that there may be many more proteins than there are genes—that is, more than 25,000. Among other advances, the database of proteins derived from proteomics is expected to help scientists better understand the regulation of gene expression in the body and how it leads to the complexity of cellular structures and functions. In addition, proteomics may lead to the development of breakthrough drugs for a variety of genetic disorders.

Genes and Our Wrld

Breakthroughs in decoding and manipulating the genetic information stored in DNA promise a world of benefits. These scientific advances already help to diagnose and treat disease, develop new medicines, bring criminals to justice, improve our food supply, and clean up the environment. At the same time, however, genetic technologies also present society with the potential for new and serious social or environmental problems. Many of the developments that worry critics of genetic technologies remain on the horizon, but the debate over their inevitable arrival is already in full swing.

Genes and the Environment

Humans have tampered with the genetic composition of other organisms for thousands of years. Most of this manipulation has been decidedly low-tech: domestication of animals and selective breeding of desirable food crops. The development and use of genetic engineering techniques has accelerated the pace at which humans can alter nature, creating some products that have unquestionable benefits and others that raise serious concerns. Some scientists fear that genetic engineering techniques will damage genetic diversity. Domestication and selective breeding, which aim to produce many similar organisms with particular desirable traits, reduce the natural variation of genes within a species. Genetic engineering techniques take this effect a giant step further. Many crop species—including wheat, corn, tomatoes, and strawberries—have been manipulated to maximize yield, appearance, resistance to pests and chemicals, hardiness, and other commercially valuable traits. Once attractive varieties have been developed, agricultural techniques favour wide-scale planting of such genetically similar or identical stocks. The impact of this practic on biodiversity can be ominous because older plant varieties that carry diverse and useful alleles may be lost forever.

The forfeited genetic material could leave a species vulnerable to annihilation from a single factor in the environment, such as an insect pest or an infectious virus. Fortunately, efforts to combat decreaing biodiversity are under way. Farmers, gardeners, government agencies, and other interested parties have collaborated to create seed banks to maintain genetic diversity. These banks catalogue, store, and distribute the seeds of rare or endangered plants, enabling gardeners and farmers to continue cultivating rare plant varieties so that the unique genetic makeup of these plants does not disappear. In the same vein, zoos and other institutions breed endangered species of animals that may no longer be able to survive in their native habitats. Other animal programmes seek to preserve or enhance the genetic variation within certain endangered animal populations. For example, all cheetahs are almost identical genetically, most likely due to their near extinction about 12,000 years ago. Inbreeding among the few remaining individuals has resulted in a loss of genetic diversity in modern cheetahs that may have affected the cheetah's immune system, leaving the animal vulnerable to disease. Scientists hope to use genetic engineering techniques to introduce new genes into the cheetah population to increase the genetic diversity of the species. The broad use of genetic engineering techniques in agriculture has raised other concerns beyond issue of biodiversity. From a consumer point of view, for instance, new technologies may have compromised a food's taste or nutritional value in exchange for a plant or animal that can be grown faster or at less cost.

In addition, some critics question the safety of genetically engineered foods. They fear that plants or animals that have received new genes will produce proteins that would not be present in nonmanipulated organisms. Such changes could have a serious effect: causing allergies or toxicity in humans who eat these foods, for example, or disrupting a plant's production of key nutrients. Though there has been much speculation about the potential health risks of genetically engineered foods, rigorous scientific investigation into the effects of these foods on humans is just beginning. The potential environmental impact of genetically engineered agriculture is equally controversial.

Critics fear that transgenic organisms, which contain DNA from other species, could give rise to populations of genetically altered life forms that could cause disease, displace native species, or otherwise harm delicate ecosystems. Consider the example of a genetically engineered form of oilseed rape, the plant that yields canola oil. The altered form, which has been grown commercially in the United States since 1993, contains inserted genes that increase its resistance to herbicides (weed-killing chemicals). The altered plants can grow unabated even when a farmer sprays a field with enough herbicide to kill troublesome weeds. Yet a curious problem has emerged: The transgenic rape can interbreed with weedy relatives that grow nearby. Some scientists fear that this interbreeding could create weed varieties that also have a genetic resistance to known herbicides. The biological and economic impact of such a development could be enormous.

Genes and Society

Advances in genetic technologies allow scientists to take an unprecedented glimpse into the genetic makeup of every person. The information derived from this testing can serve many valuable purposes: It can save lives, assist couples trying to decide whether or not to have children, and help law-enforcement officials solve a crime.Yet breakthroughs in genetic testing also raise some troubling social concerns about privacy and discrimination. For example, if an individual's genetic information becomes widely available, it could give health insurers cause to deny coverage to people with certain risk factors or encourage employers to reject certain high-risk job applicants. Furthermore, many genetically linked problems are more common among certain racial and ethnic groups—for example, the BRCA1 breast cancer allele is more common in Ashkenazi Jews, and the blood disorder sickle-cell anemia is more prevalent among blacks of African ancestry.

Many minority groups fear that the expansion of genetic testing could create whole new avenues of discrimination. Of particular concern are genetic tests that shed light on traits such as personality, intelligence, and mental health or potential abilities. Genetic tests that indicate a person is unlikely to get along with other people could be used to limit a person's professional advancement. In other cases, tests that identify a genetic risk of heart failure could discourage a person from competing in sports. New technologies that allow the manipulation of genes have raised even more disturbing possibilities. Gene therapy advances, which allow scientists to replace defective genes with normal alleles, give people with typically fatal diseases new hope for healthy lives. To date, gene therapy has focused on manipulating the genetic material in body cells other than gametes, so the changes will not be passed on to future generations. However, the application of gene therapy techniques to gametes—the cells involved in reproduction—seems inevitable.

Such manipulation might help prevent the transmission of disease from one generation to another, but it could also produce unforeseen problems with long-lasting consequences. For instance, many people worry that new genetic techniques could be used to alter or encourage traits now viewed as part of normal human variability, such as shortness or baldness. At various times in the past century, people have advocated efforts to improve the human condition by promoting the perpetuation of certain genes. This concept, known as eugenics, typically involves encouraging people with "positive" genes to reproduce and discouraging those with "inferior" genes from having offspring. Many people fear that new genetic technologies used to manipulate the human genome could give people previously unattainable methods to resort to extreme forms of eugenics. Advances in genetic technologies have turned some genes into valuable commercial commodities, spawning a host of controversial questions. Who owns a genetically altered organism or the genes it contains? Is it right to patent the use of a naturally occurring gene? Some people feel that genetic material should not be owned or used for profit.

Costa Rica has enacted laws to prevent foreign companies from patenting and then profiting from genes of native Costa Rican plant and animal species. Balancing the need to limit patents on genes are concerns that the profit motive of companies must be protected to maintain incentives to make new discoveries for medical products. The citizens of Iceland, for example, are cooperating with a biotechnology company in a study of the genetic makeup of the Icelandic people. The information compiled will be used to learn about genetic diseases.

Heredity

Humans have had some understanding of heredity since prehistoric times, observing how similar traits pass from parent to offspring and noting that differences arise with each generation. Most of the mechanisms of heredity, however, were shrouded in mystery until early in the 20th century. Since that time, the rate of discovery has reached a feverish pace, enabling the advancement of modern molecular biology and the current Human Genome Project.

Early Views of Heredity

In ancient times, people understood some basic rules of heredity and used this knowledge to breed domestic animals and crops. By about 5000 BC, for example, people in different parts of the world had begun applying selective breeding techniques to grow new plant varieties, including types of wheat, maize, rice, and date palms, that had never existed in the wild. Ancient people understood that the rules of inheritance also applied to humans. The ancient Greeks were particularly interested in human heredity and evolution. Greek scientists and philosophers hotly debated whether a male or female parent contributed more to an offspring.

In the 4th century BC, Aristotle speculated that acquired characteristics, such as a scar that was incurred during life, could be passed on to offspring. He also believed in a widely held theory known as pangenesis. This theory proposed that particles in the body, called gemmules, reside in the limbs and organs. The gemmules become imprinted with any changes acquired by the body, such as muscle development from exercise. The gemmules then move to the reproductive cells and transfer information about the body's alterations to these cells. The reproductive cells transmit the acquired traits to offspring through particles called pangenes. The theories about the inheritance of acquired characteristics and pangenesis persisted until the middle of the 19th century. French zoologist Jean-Baptiste Lamarck formalized the theory of acquired characteristics in his treaise *Philosophie Zoologique*. Lamarck proposed that organisms evolve by responding to changes in their environment. When organisms undergo a change in order to adjust to their environment, that change acts as a trait that can be passed on to offsprng.

Influences of Darwin and Mendel

A surprising supporter of pangenesis was the British naturalist Charles Robert Darwin, who believed that the theory accounted for the process of heredity and the wide variety of traits seen among offspring. Despite his mistaken belief in pangenesis, Darwin nonetheless had an enormous impact on human understanding of heredity. During his years of extensive worldwide travel, Darwin collected many observations of how related species adapt to their local environments.

Darwin and British naturalist Alfred Wallace independently formulated the theory of natural selection, which holds that members of a given species born with more favorable characteristics to deal with their environment would be most likely to survive to pass on these traits to the next generation. This important theory was popularized by Darwin's publication *On the Origin of Species*. The book was an immediate sensation, but it raised many questions. Foremost among these was the mystery of how organisms could appear with modified or entirely new traits.

At roughly the same time that Darwin published his natural selection theories, the answer to many questions about the mechanisms of heredity were being unraveled by Gregor Mendel, a reclusive Austrian monk. Mendel conducted a long series of experiments on pea plants during the 1850s and 1860s. Mendel crossbred plants that expressed differing traits, such as height and flower colour. His conclusions from these experiments helped him formulate a comprehensive theory of how such traits pass from one generation to another. In his studies, Mendel recognized that characteristics were inherited as discrete units, and that each of these was inherited independently of the others. He speculated that each parent has pairs of these units but passes only one to an offspring. He also noted that certain forms of one trait were always dominant over others. Today the units that Mendel described are known as genes.

Emergence of the Science of Genetics

Mendel published his findings in 1866, but they went largely unnoticed for more than three decades. In the year 1900, however, Dutch botanist Hugo Marie de Vries, German botanist Karl Correns, and Austrian botanist Erich Tschermak independently rediscovered the monk's works and verified his conclusions. Advances in cytology, the science of the structure and function of cells, enabled scientists to more deeply appreciate Mendel's work. In 1902 American biologist Walter S. Sutton and German cell biologist Theodor Boveri separately noted the parallels between Mendel's units and chromosomes. The demonstration of the chromosomal basis of inheritance gave rise to the modern science of genetics. The term *genetics* itself was coined in 1905 by British biologist William Bateson. The terms *gene* and *genotype* were contributed in 1909 by German scientist Wilhelm Johannsen.

In 1905 American biologists Edmund B. Wilson and Nettie Stevens independently discovered and identified the sex chromosomes. Wilson discovered the X chromosome in a butterfly, and Stevens discovered the Y chromosome in a beetle. The discoveries of the X and Y chromosomes helped scientists begin to unravelnew patterns of inheritance. Foremost among this research was the work of American biologist Thomas Hunt Morgan on fruit flies. In 1910 Morgan identified the first proof of a sex-linked trait, an eye-colour characteristic that resides on the X chromosome of fruit flies. With this finding, Morgan became the first scientist to pin down the location of a gene to a specific chromosome. Morgan was also the first to explain the implications of linkage, unuual patterns of inheritance that occur when multiple genes found on the same chromosome are inherited together. A student of Morgan's, American biologist Alfred Sturtevant, found early evidence of the mechanisms of crossing over, the phenomenon in which chromosomes interchange genes. More definitive proof emerged in the 1930s with work by American geneticists Harriet Creighton and Barbara McClintock. The pair demonstrated gene recombination with experiments on seed colour in corn. McClintock later gained notice for her work on transposable elements, large genetic segments that move within a chromosome or even between chromosomes. Her research into these elements, commonly known as jumping genes, earned McClintock the 1983 Nobel Prize in physiology or mediine.

Breakthroughs in DNA Studies

While cytologists and geneticists were studying the properties and location of genes on chromosomes, other scientists focused their studies on the composition of genes. In 1928 British microbiologist Frederick Griffith ran a series of experiments on two strains of bacteria, one that kills mice and another that is harmless to them. When Griffith injected mice with killed cells of the virulent bacteria, all of the mice survived. But in a second trial, when Griffith injected a combined cocktail of dead virulent bacteria and live "harmless" bacteria, the mice all died. He concluded that something in the dead virulent cells "transformed" the hereditary material of normally harmless bacteria so that they became killers. Most scientists at the time theorized that the transforming factor was composed of a protein.

The real identity of the transforming factor in this experiment was not identified until 1944, when American geneticists Oswald Avery, Colin MacLeod, and Maclyn McCarty revisited Griffith's research. After isolating different molecular components from dead bacterial cells, Avery and his colleagues determined that DNA was the agent that transformed the live harmless bacteria into killers. Despite a growing body of evidence about the function of DNA, many scientists were not ready to reject proteins as the hereditary material. The debate was largely quieted in 1952 by American geneticists Alfred Hershey and Martha Chase. Hershey and Chase showed that when a type of virus called a bacteriophage infects a bacterium, it is the virus's DNA—not protein—that enters the bacterium to cause infection. Their studies confirmed that DNA contained the virus's genetic information, which triggered vral replication within the bacteria. The experiments of Hershey and Chase convinced most scientists that DNA was the molecule of heredity, but many questions about the structure and mechanisms of DNA remained. In the early 1950s researchers began to apply techniques of X-ray diffraction to learn about the basic structure of DNA. X-ray diffraction can determine molecular structures by measuring patterns of scattered X rays after they pass through a crystalline substance. British physical chemist Rosalind Franklin and British biophysicist Maurice Wilkins used X-ray diffraction to obtain DNA images of unprecedented clarity. Yet the exact three-dimensional structure of DNA remained unclear. The groundbreakig work of American biochemist James Watson and British biophysicist Francis Crick solved that mystery. In 1953 the two proposed a model of DNA that is still accepted today: A double helix molecule formed by two chains, each composed of alternating sugar and phosphate groups, connected by nitrogenous bases. Watson and Crick were awarded the 1962 Nobel Prize in physiology or medicine for their discoveries.

Watson and Crick speculated that the structure of DNA provided some obvious clues about how the molecule could replicate itself. They proposd a replication model in which each strand of DNA serves as a template for making exact copies. This mode of replication, called semi-conservaive replication, was demonstrated in 1958 by American molecular biologists Matthew Meselson and Franklin Stahl. Their experiments demonstrated the mechanism of replication by tracking DNA containing a heavy nitrogen isotope through a serie of replications. With DNA's structureand replication mechanisms largely solved, scientists turned their attention to identifying the genetic code—learning how a gene's nucleotide sequence determines what type of protein is made. In the late 1950s, South African geneticist Sydney Brenner and other scientists confirmed that RNA acted as an intermediary between DNA and protein production.

Researchers still were uncertain how the sequence of nucleotides in DNA corresponded to the production of specific amino acids. In 1961 Crick and Brenner determined that groups of three nucleotides, now known as codons, code for the 20 amino acids that form the foundation of proteins. The exact relationship between codons and amino acids was clarified after several important discoveries. American biochemists Marshall Nirenberg and J. Heinrich Matthaei synthesized repeated nucleotide sequences that led to the production of repeated single amino acids. They identified how certain codon combinations code for a specific amino acid. A process developed by American geneticist Har Gobind Khorana helped scientists create a "dictionary" of codons that defined specific amino acids, thus resolving the remaining ambiguities in the genetic code. Only 12 years after the structure of DNA was deduced, the genetic code was solved.

Learning to Manipulate DNA

After scientists had unraveled the structureand replication mechanisms of DNA, many felt that the major discoveries of genetic research were resolved. They predicted that the only task left in genetics was to sort out the molecular details of how genes work. But in the process of studying gene function, researchers developed powerful new molecular techniques, enabling them to analyse and manipulate genes with a speed and precision never before possible. A number of discoveries made during the 1960s and 1970s shed light on how distinct fragments of DNA could be isolated. The work of Swiss molecular biologist Werner Arber focused on specialized enzymes that digest, or "restrict," the DNA of viruses infecting bacteria. These enzymes were subsequently dubed restriction enzymes. In the following decade, scientists learned hat restriction enzymes could also act like molecular scissors to cut DNA.

In 1970 American molecular biologist Hamilton Smith and colleagues determined hat restriction enzymes could cleave DNA molecules at precise and predictable locations. Hamilton concluded that the enzymes were able to recognize specific nucleotide sequences. Scientists quickly realised hat restriction enzymes could be used in the laboratory to manipulate DNA. In 1973 American biochemist Herb Boyer sed restriction enzymes to produce a DNA molecule with genetic material from two different sources. This splicing technique is now know as recombinant DNA. Boyer inserted foreign genes into plasmids and observed that the plasmids could replicate to make many copies of the inserted genes. In subsequent experiments, Boyer, American biochemist Stanley Cohen, and other researchers demonstrated that insertig a recombinant DNA molecule into a host bacteria cell would lead to extremely rpid replication and the production of many identical copies ofthe recombinant DNA.

This process, known as cloning, gave scientists the power to make many copies of desired DNA for molecular study. The speed and efficiency of DNA cloning were vastly improved in the 1980s with the invention of polymerase chain reaction (PCR). Developed by American biochemist Kary Mullis, PCR enables scientists to produce large amounts of DNA sequences in a test tube. In a matter of hours, the process can produce millions of cloned DNA molecules. Yet all of the advances in isolating and replicating DNA would not be possible or be of much use if researchers could not determine the nucleotide sequence of genetic material.

In the late 1970s and early 1980s, British biochemist Frederick Sanger and his associates developed DNA sequencing techniques. Sanger's methods, which used special compounds called dideoxy nucleotides, rapidly yielded the exact nucleotide sequence of a desired sample. With the use of automated equipment, the new techniques transformed genetic sequencing into a speedy, routine laboraory procedure. Many of the new techniques for isolating, sequencing, and replicating DNA have been put to practical use through the field of genetic engineering. The Human Genome Project and the new field of proteomics have both benefited from continuing technical advances and have accelerated the development of new genetic technologies. Modern genetics is poised to radically change the practice of medicine and the biotechnology industry.

Cutting and Measuring DNA

The first step in gene cloning is cutting DNA into appropriately sized pieces. The size of DNA molecules is measured by subjecting the DNA to a technique called "electrophoresis". An electrophoretic apparatus gives DNA an electrical shove (DNA is negatively charged because of its phosphate groups. The electrophoresis apparatus puts the DNA in an electric field, and it moves towards the positive pole), forcing it to move through a porous gel. Two types of gels are commonly u.

Agarose Gels

The most popular variant of electrophoresis is carried out th agarose gels. In the laboratory, suspensions of agarose in buffer are heated to boiling (often in microwave ovens), and the hot solution is poured on plastic or glass plates to a height of a few millimeters or so. Agarose is a derivative of agar — an edible polysaccharide extracted from seaweed. When cooled from heated solutions, solutions of agarose form a nearly transparent gel that looks like a slab of gelatin dessert. After the gel sets, DNA solutions are placed in little depressions formed by leaving a comb in the gel. The gel is then flooded with a weak salt solution, and a voltage is applied. Since DNA is negatively charged (due to its phosphate groups), it moves towards the positive pole. Large DNA molecules advance slowly in the gel because they are impeded by the gel matrix.

Smaller ones encounter fewer barriers and move more quickly. By plotting the distance migrated against the reciprocal of size (in base pairs) of a group of DNA standards, a fairly straight line can be obtained. From these data, the length of unknowns can be easily calculated. For ease of comparison, the standards are usually run in the same gel beside the unknowns. Gels of differing porosities can be made by adjusting the concentration of agarose. With 2 and 3 per cent solutions, double-stranded DNA's as small as 50 or 100 base pairs can be resolved. More dilute gels can resolve fragments as large as about 30,000 base pairs (30 kilobase pairs or 30kb).

Polyacrylamide Gels

When very small DNA molcculcs need to be analysed or when high-resolution separations are required, electrophoresis is carried out in polyacrylamide gels. They, like the gels made from agarose, are water white. Unlike agarose, polyacrylamide gels arc polymers of acrylamide, a small synthetic organic compound. Polyacrylamide gel electrophoresis (PAGE) is most often used to separate fragments of DNA that are between 6 and a few thousand base pairs in length. The technique is used mostly in DNA sequencing, a topic that we'll come back to later in the course.

Detecting DNA

DNA doesn't have any colour; it only absorbs light in the ultraviolet range. How is it visualized in a clear gel after electrophoresis? DNA can be detected in both agarose and polyacrylamide gels after staining with various dyes, incuding ethidium bromide, a dye that forms a fluorescent complex upon binding to DNA. Usually, the ethidium bromide is added to the gel before electrophoresis. After the run, the gel is examined and often photographed under an ultraviolet light. Yellow-orange zones (bands) of fluorescence appear (similar to those in the picture at the right), indicating migration of discrete pieces of DNA. Other agents — like methylene blue or silver stains — can also be used to visualize DNA, but they are either less sensitive or less convient, and aren't used as often. By the way, many dyes that bind to DNA are mutagenic; i.e, they cause mutations.

Separating Large Fragments of DNA

Contrary to what we might expect, very large fragments of DNA readily movin agarose gels when an electrical field is applied. But they don't separate according to size. In fact, very big pieces of DNA all migrate at about the same velocity regardless of whether they are 20kb in length or 2000kb.

PFE

Since many techniques, particulary recently derived ones, depend on working with very large DNA fragments, even pieces the size of some chromosomes, several newly developed electrophoretic techniques have been developed. In one of these, called pulsed field electrophoresis (PFE), DNA molecules are analyseon agarose gels but with a modified apparatus. Instead of having only two sets of electrodes situated at opposite ends of the electrophoresis device as in the picture shown above, the pulsed field apparatus bears four sets of electrodes.

Two are designated A^- and A^+ and the other two are called B^- and B^+. The two A's and the two B's are set at an angle of 120° with respect to one another. During electrophoresis, the DNA molecules are subjected to alternating bursts of current from the two A and B pairs, thereby pulling the DNA alternately to the right and to the left as it advances (from south to north in this case) through the gel. The DNA seems to reorient itself each time the current is switched, and the time that it takes to turn itself around is dependent on its length. The result is that very large pieces of DNA can be separated from another on the basis of their size.

The technique also allows one to estimate the length of an unknown when run beside standards of known size. One of the most impresive accomplishments of this technique is the separation of the 16 chromosomes of yeast in a single electrophoretic run.

FIE

Anoher procedure, field inversion electrophoresis, is also effective at separating large molecules of DNA. In it, the electrodes are set up as in ordinary electrophoresis, but the current is switched so that the DNA first moves forward and then backward. Of course, if the switching were to be done in equal time intervals — say, 1 second forward and 1 second back — the DNA wouldn't move at all. To get it to move at all, electrophoresis is conducted forward for a longer time than backward (or at a greater voltage forward versus backward). One might expect that moving ahead three steps and back two would be equivalent to moving forward one step at a time, but that's not what happens. Apparently the DNA reorients itself during the switching cycles, just as it does in pulsed field electrophoresis. Again, the ability to change directions seems to be dependent on size. This allows different lengths of DNA to be separated from one anoter.

Restriction Endonucleases

One critically important advance that has greatly stimulated the rapid progress in molecular biology and genetic engineering was the discovery of a set of enzymes that are capable of cutting DNA at defined sequences. These enzymes are found in a variety of microorganisms and are caled restriction endonucleases, or more simly, restriction enzymes. The first specific restriction enzyme was discovered by Hamilton Smith in 1970, an accomplishment for which he (and two others) were awarded the Nobel Prize. Since then, some 3,059 similar enzymes have been reported, many of which have different specificites.

Nomenclature

Smith and another Nobel laureate, Daniel Nathans, devisd a nomenclature for these enzymes. In brief, the name of each restriction enzyme derives from the organism from which it is isolated. The first letter of the genus name plus the first two letters of the species name form the first three letters of the restriction enzyme's name. If necessary, a letter indicating strain designation is added, and finally a number is appended that stands for the order in which the enzyme was discovered in each organism.

For example, BamHI is the name of an enzyme that is isolated from the bacteium *Bacillus amyloliquifaciens,* strain H, and it was, presumably, the first restriction endonuclease identified from that source. The first three letters of the name should be italicized, but because italicized letters are often hard to read on a computer.

Restriction Enzymes

The restriction enzymes owe their usefulness to the fact that they bind to DNA at specific DNA sequences, four to eight nucleotides in size, called recognition sites. Once bond, restriction enzymes cut the DNA at or near this site. With a little thought, it should be clear that an enzyme that has a six base pair recognition site will, on the average, produce larger pieces of DNA than one that recognizes a four base site. Expressed quantitatively, the approximate size of the fragments produced by a particular enzyme, given that it is cutting a DNA containing an equal proportion of all four nucleotides, can be calculated from the formula:

$$\text{Average size of fragment} = 4^N$$

where N is the number of bases that the enzyme recognizes. Hence, a four-cutter (the shorthand name for an enzyme that recognizes a site containing four base pairs) is expected to cleave random DNA into fragments of about 4^4 (256) base pairs while an enzyme with a recognition site of six bases will produce pieces (on the average) of about 4^6 (4096) base pairs.

Propertie of Restriction Enzymes

Ends

Another interesting propert of restriction enzymes is that while they often recognize a symmetrical site, they do not always cut at the axis of symmetry. For instance, the enzyme EcoRI. Similarly, the enzyme BglII (from the microorganims, *Bacillus globiggi* and universally and irreverently pronounced BAGEL TWO) recognizes the sequence AGATCT and cuts between the first A and G residues. In fact, the four nucleotide single-stranded ends are the same for both BglII and BamHI. Moreover, there are at least two other six cutting enzymes that have been discovered that leave the same four nucleotide overhang: BclI and XhoII. These overhanging ends are very useful because — under the proper conditions — they may base pair with each other. In fact, because of their affinity for one another, they are often called cohesive or sticky ends. Moreover, if molecules with these ends are treated with the appropriate enzyme — DNA ligase — their phosphodiester bonds may be rejoined (ligated). When two ends that originate from digestion by a single enzyme are ligated, the resulting molecule can be cut by the same enzyme again. But if the ends of a DNA molecule that originated with a BamHI cut and a BglII cut are joined together, the new sequence will not be cut with either enzyme.

All restriction endonucleases do not generate 5' single-strand overhangs. In fact, some don't even produce an overhang at all. Several enzymes — like SacI — produce 3' single-stranded sticky ends. And some enzymes — like PvuII — cut at the axis of symmetry, leaving perfectly aligned ends. DNA molecules without overhangs are said to have blunt ends.In addition thereare restriction enzymes that cleave DNA some distance away from the sequence that they recognize. For example the enzyme HgaI makes staggered cuts that lie 5 and 10 nucleotides away from a 5 base pair sequence, GACGC. This leaves 5' overhanging ends, but, in contrast to the enzymes described above, these will be different almost every time the enzyme cuts.

The Utility ofthe Restriction Enzymes

The discovery of these any restriction endonucleases have allowed genetic engineers to cut pieces of DNA at specific sites and into defined sizes. The result has been that a scientist can work with a collection of molecules all of the same size and with ends of known sequece. Restriction enzymes have proved to be valuable analytical and diagnostic tools as wll.

Restriction Enzmes

Restriction enzymes are DNA-cutting enzymes found in bacteria (and harvested from them for use). Because they cut within the molecule, they are often caled restriction endonucleases. A restriction enzyme recognizes and cuts DNA only at a particular sequence of nucleotides. For example, the bacterium Hemophilus aegypticus produces an enzyme named HaeIII that cuts DNA wherever it encounters the sequence,

5′GGCC3'

3′CCGG5'

The cut is made between the adjacent G and C. This particular sequence occurs at 11 places in the circular DNA molecule of the virus phiX174. Thus treatment of this DNA with the enzyme produces 11 fragments, each with a precise length and nucleotide sequence. These fragments can be separated from one another and the sequence of each determined. HaeIII and AluI cut straight across the double helix producing "blunt" ends. However, any restriction enzymes cut in an offset fashion. The ends of the cut have an overhanging piece of single-stranded DNA. These are called "sticky ends" because they are able to form base pairs with any DNA molecule that contains the complementary sticky end. Any other source of DNA treated with the same enzyme will produce such molecules.

Mixed together, these molecules can join with each other by the base pairing between their sticky ends. The union can be made permanent by another enzyme, DNA ligase, that forms covalent bonds along the backbone of each strand. The result is a molecul of recombinant DNA (rDNA). The ability to prouce recombinant DNA molecules has not only revolutionized the study of genetics, but has laid the foundation for much of the biotechnology industry. The availability of human insulin (for diabetics), human factor VIII (for males with hemophilia A), and other proteins used in human therapy all were made possibl by recombinant NA.

Recombinant DNA and Gene Cloing

Recombinant DNA is DNA that has been created artificially. DNA from two or more sources is incorporated into a sigle recombinant molecule.

Maing Recombinant DNA (rDNA)

- Treat DNA from both sources with the same restriction endonuclease (BamHI in this case).
- BamHI cuts the same site on both molecules

  5' GGATCC 3'

  3' CCTAGG 5'
- The ends of the cut have an overhanging piece of single-stranded DNA.
- These are called "sticky ends" because they are able to base pair with any DNA molecule containing the complementary sticky end.
- In this case, both DNA preparations have complementary sticky ends and thus can pair with each other when mixed.
- DNA ligase covalently links the two into a molecul of recombinant DNA.

To be useful,the recombinant molecule must be replicated many times to provide material for analysis, sequencing, etc. Producing many identical copies of the ame recombinant molecule is called cloning. Cloning can be done in vitro, by a process called the polymerase chain reaction (PCR). Here, however, we shall examine how cloning is done in vivo.

Cloning in vivo can be done in:

- Unicellular microbes like E. coli
- Unicelllar eukaryotes like yeast and
- In mammalian cells grown in tissue culture.

In every case,the recombinant DNA must be taken up by the cell in a form in which it can be replicated and expressed. This is achieved by incorporating the DNA in a vector. A number of viruses (both bacterial and of mammalian cells) can serv as vectors. But here let us examine an example of cloning using E. coli as the host and a plasmid as the vector.

Plasmid

A plasmid is an extra-chromosomal DNA molecule separate from the chromosomal DNA which is capable of replicating independently of the chromosomal DNA. In many cases, it is circular and double-stranded. Plasmids usually occur naturally in bacteria, but are sometimes found in eukaryotic organisms (*e.g.*, the 2-micrometre-ring in Saccharomyces cerevisiae). Plasmid size varies from 1 to over 200 kilobase pairs (kbp). The number of identical plasmids within a single cell can be zero, one, or even thousands under some circumstances. Plasmids can be considered to be part of the mobilome, since they are often associated with conjugation, a mechanism of horizontal gene transfer.The term plasmid was first introduced by the American molecular biologist Joshua Lederberg in 1952.

Plasmids can be considered to be independent life-forms similar to viruses, since both are capable of autonmous replication in suitable (host) environments. However the plasmid-host relationship tends to be more symbiotic than parasitic (although this can also occur for viruses, for example with Endoviruses) since plasmids can endow their hosts with useful packages of DNA to assist mutual survival in times of severe stress. For example, plasmids can convey antibiotic resistance to host bacteria, who may then survive along with their life-saving guests who are carried along into future host generatons.

Vectors

There are two types of plasmid integration into a host bacteria: Non-integrating plasmids replicate as in the top instance; whereas episomes, the lower example, integrate into the host chromosome. Plasmids used in genetic engineering are clled vectors. Plasmids serve as important tools in genetics and biotechnology labs, where they are commonly used to multiply (make many copies of) or express particular genes. Many plasmids are commercially available for such uses. The gene to be replicated is inserted into copies of a plasmid containing genes that make cells resistant to particular antibiotics and a multiple cloning site (MCS, or polylinker), which is a short region containing several commonly used restriction sites allowing the easy insertion of DNA fragments at this location.

Next, the plasmids are inserted into bacteria by a process called transformation. Then, the bacteria are exposed to the particular antibiotics. Only bacteria which take up copies of the plasmid survive the antibiotic, since the plasmid makes them resistant. In particular, the protecting genes are expressed (used to make a protein) and the expressed protein breaks down the antibiotics. In this way the antibiotics act as a filter to select only the modified bacteria. Now these bacteria can be grown in large amounts, harvested and lysed (often using the alkaline lysis method) to isolate the plasmid of interest. Another major use of plasmids is to make large amounts of proteins. In this case, researchers grow bacteria containing a plasmid harboring the gene of interest. Just as the bacteria produces proteins to confer its antibiotic resistance, it can also be induced to produce large amounts of proteins from the inserted gene. This is a cheap and easy way of mass-producing a gene or the protein it then codes for, for example, insulin or even antibiotics. However, a plasmid can only contain inserts of about 1-10 kbp. To clone longer lengths of DNA, lambda phage with lysogeny genes deleted, cosmids, bacterial artificial chromosomes or yeast artificial chromosomes could be used.

Types

One way of grouping plasmids is by their ability to transfer to other bacteria. Conjugative plasmids contain so-called tra-genes, which perform the complex process of conjugation, the transfer of plasmids to another bacterium. Non-conjugative plasmids are incapable of initiating conjugation, hence they can only be transferred with the assistance of conjugative plasmids, by 'accident'. An intermediate class of plasmids are mobilizable, and carry only a subset of the genes required for transfer. They can 'parasitize' a conjugative plasmid, transferring at high frequency only in its presence. Plasmids are now being used to manipulate DNA and may possibly be a tool for curing many diseases. It is possible for plasmids of different types to coexist in a single cell. Seven different plasmids have been found in E. coli. But related plasmids are often incompatible, in the sense that only one of them survives in the cell line, due the regulation of vital plasmid functions.

Therefore, plasmids can be assigned into compatibility groups. Another way to classify plasmids is by function.

There are five main classes:

1. Fertility-F-plasmids, which contain tra-genes. They are capable of conjugation.
2. Resistance-(R)plasmids, which contain genes that can build a resistance against antibiotics or poisons. Historically known as R-factors, before the nature of plasmids was understood.
3. Col-plasmids, which contain genes that code for (determine the production of) bacteriocins, proteins that can kill other bacteria.
4. Degradative plasmids, which enable the digestion of unusual substances, *e.g.*, toluene or salicylic acid.
5. Virulence plasmids, which turn the bacterium into a pathogen.

Plasmids can belong to more than one of these functional groups. Plasmids that exist only as one or a few copies in each bacterium are, upon cell division, in danger of being lost in one of the segregating bacteria. Such single-copy plasmids have systems which attempt to actively distribute a copy to both daughter cells. Some plasmids include an addiction system or "postsegregational killing system (PSK)", such as the hok/sok (host killing/suppressor of killing) system of plasmid R in Escherichia coli. They produce both a long-lived poison and a short-lived antidote. Daughter cells that retain a copy of the plasmid survive, while a daughter cell that fails to inherit the plasmid dies or suffers a reduced growth-rate because of the lingering poison from the parent cell.

Plasmid DNA Extraction

As alluded to above, plasmids are often used to purify a specific sequence, since they can easily be purified away from the rest of the genome. For their us as vectors, and for molecular cloning, plasmids often need to be isolated. There are several methods to isolate plasmid DNA from bacteria, the archetypes of which are the miniprep and the maxiprep/bulkprep. The former can be used to quickly find out whether the plasmid is correct in any of several bacterial clones. The yield is a small amount of impure plasmid DNA, which is sufficient for analysis by restriction digest and for some cloning techniques. In the latter, much larger volumes of bacterial suspension are grown from which a maxi-prep can be performed. Essentially this is a scaled-up miniprep followed by additional purification. This results in relatively large amounts (several micrograms) of very pure plasmid DNA

DNA Ligases

DNA ligases are vital enzymes required for important cellular processes such asDNA replication, repair of damaged DNA and recombination. The enzyme mediates the formation of phosphodiester bonds between adjacent 3'-OH and 5'-phosphate termini, thereby joining the nicks in double stranded DNA. Ligases can be classified into two groups depending on their requirement for ATP or NAD+ as the cofactor. All eukaryotic and virally encoded enzymes are ATP-dependent, whereas most prokaryotic enzymes require NAD^+ for their activity.

ATP-Dependent DNA Ligase

DNA ligase from bacteriophage T7 is a monomer with a molecular weight of 41 kDa. Crystal structures of the apo form of this enzyme and its complex with ATP have been determined at 2.7 Å and 2.6 Å respectively. The structure consists of two distinct domains, a larger N-terminal domain and a C-terminal domain. The ATP-binding site is situated in the N-terminal domain in a pocket beneath one of the beta-sheets. This pocket is lined by a number of motifs that are conserved across a wide family of nucleotidyltransferases.

The structure of the C-terminal domain is remarkably similar to the oligonucleotide binding fold, observed in a number of proteins. The DNA-binding site is proposed to be in a groove running between the two domains. The structure has provided the basis for biochemical experiments. We have made the two domains separately and looked at their biochemical properties. The larger N-terminal domain is an active ligase but with greatly reduced activity. Both domains are able to bind to DNA. The N-terminal domain binds both single and double stranded DNA, while the smaller C-terminal domain is only able to bind double stranded DNA despite having a fold that is similar to single strand DNA binding proteins.

The affinity for nicked DNA comes from a combination of these two DNA-binding affinities at the active site. Interestingly, the N-terminal domain is poor at the adenylation reaction, but this activity is stimulated greatly by addition of the C-terminal domain. Furthermore, we can demonstrate an association of the domains on gel filtration. These data suggest that a conformational change occurs during the adenylation reaction. This would be similar to that observed directly in our crystal structure of an mRNA capping enzyme, a group of enzymes that are closely related to ligases.

NAD-Dependent DNA Ligase

An NAD-dependent DNA ligase from B.stearother-mophilus. The enzyme has been overexpressed and biochemical studies are in progress. Limited proteolysis has been used to investigate the domain structure of the enzyme. There are two domains, with the larger N-terminal domain retaining full self adenylation activity and the smaller C-terminal domain having all of the DNA-binding activity of the full length enzyme. The situation is therefore different to that we found in the ATP-dependent enzyme, where both domains contribute to the activities of the enzyme. Structural studies are underway with these domains. The structure of the large N-terminal domain has been solved and work is in progress to determine the structure of the small domain.

Ribonuclease A

Ribonuclease A (RNase A) is an endonuclease that cleaves single-stranded RNA. Bovine pancreatic RNase A is one of the classic model systems of protein science. The importance of bovine pancreatic RNase A was secured when the Armour purifid a *kilogram* of it, and gave 10 mg samples away free to any interested scientists. The ability to have a single lot of purified enzyme instantly made RNase *the* model system for protein studies. RNase A was the model protein used to work out many spectroscopic methods for assaying protein structure, including absorbance, circular dichroism/optical rotary dispersion, Raman, EPR and NMR spectroscopy. RNase A was also the first model protein for the development of several chemical structural methods, such as limited proteolysis of disordered segments, chemical modification of exposed side chains, and antigenic recognition.

Ribonuclease-S, which is RNase A that has been treated with subtilisin, was the third protein to have its structure solved, in 1967.Studies of the oxidative folding of RNase A led Chris Anfinsen to enunciate the thermodynamic hypothesis of protein folding, which states that the folded form of a protein represents the minimum of its free energy. RNase A was the first protein for showing the effects of non-native isomers of X-Pro peptide bonds in protein folding.RNase A was the first protein to be studied by multiple sequence alignment and by comparing the properties of evolutionarily related proteins.

Structure and Properties

Labeled ribbon diagram of bovine pancreatic ribonuclease A (PDB accesion code 7RSA). The backbone ribbon is coloured from blue (N-terminus) to red (C-terminus). The side chains of the four disulfide-bonded cysteines are shown in yellow, with their sulpher atoms highlighted as small spheres. Residues important for catalysis are shown in magenta. RNase A is a relatively small protein (124 residues, ~13.7 kDa). It can be characterized as a two-layer á + â protein that is folded in half to resemble a taco, with a deep cleft for binding the RNA substrate. The first layer is composed of three alpha helices from the N-terminal half of the protein. The second layer consist of three â-hairpins arranged in two a-sheets. The hairpins 61-74 and 105-124 form a four-stranded, antiparallel â-sheet that lies on helix 3 (residues 50-60). The longest â-hairpin 79-104 mates with a short â-strand (residues 42-45) to form a three-stranded, antiparallel â-sheet that lies on helix 2 (residues 24-34). RNase A has four disulfide bonds in its native state: Cys26-Cys84, Cys58-110, Cys40-95 and Cys65-72. The first two (26-84 and 58-110) are essential for conformational folding; each joins an alpha helix of the first layer to a beta sheet of the second layer, forming a small hyrophobic core in its vicinity. The latter two disulfide bonds (40-95 and 65-72) are less essential for folding; either one can be reduced (but not both) without affecting the native structure under physiological conditions.

These disulfide bonds connect loop segments and are relatively exposed to solvent. Interestingly, the 65-72 disulfide bond has an extraordinarily high propensity to form, significantly more than would be expected from its loop entropy, both as a peptide and in the full-length protein. This suggests that the 61-74 α-hairpin has a high propensity to fold conformationally. RNase A is a basic protein (pI =8.63); its many positive charges are consistent with its binding to RNA (a poly-anion). More generally, RNase A is unusually polar or, rather, unusually lacking in hydrophobic groups, especially aliphatic ones. This may account for its need of four disulfide bonds to stabilize its structure. The low hydrophobic content may also serve to reuce the physical repulsion between highly charged groups (its own and those of its substrate RNA) and regions of low dielectric constant (the nonpolar residues). The N-terminal a-helix of RNase A (residues 3-13) is connected to the rest of RNase A by a flexible linker (residues 16-23). As shown by F. M. Richards, this linker may be cleaved by subtilisin between residues 20 and 21 without causing the N-terminal helix to dissociate from the rest of RNase A. The peptide-protein complex is called RNase S, the peptide (residues 1-20) is called the S-peptide and the remainder (residues 21-124) is called the S-protein.

The dissociation constant of the S-peptide for the S-protein is roughly 30 pM; this tight binding can be exploited for protein purification by attaching the S-peptide to the protein of interest and passing a mixture over an affinity column with bound S-protein. The RNase S model system has also been used fr studying protein folding by coupling folding and association. The S-peptide was the first peptide from a native protein shown to have (flickering) secondary structure in isolation

Enzymatic Mechanism

The positive charges of RNase A lie mainly in a deep cleft between two lobes. The RNA substrate lies in this cleft and is cleaved by two catalytic histidines, His12 and His119, via a 2'–3' cyclic phosphate intermediate that is stabilized by nearby lysines such as Lys7, Lys41 and Lys66.

Anti-Cancer Effects

RNase A, and to a greater extent its oligomers and some homologs (such as onconase from frogs), have cytotoxic and cytostatic effects, particularly on cancer cells. This has led to the development of onconase as a cancer therapeutic, particularly for external use against skin cancers. As with many protein drugs, the internal use of non-human ribonucleases such as onconase is limited by the patient's immune response.

Heredity

Genetics, study of the function and behaviour of genes. Genes are bits of biochemical instructions found inside the cells of every organism from bacteria to humans. Offspring receive a mixture of genetic information from both parents. This process contributes to the great variation of traits that we see in nature, such as the colour of a flower's petals, the markings on a butterfly's wings, or such human behavioral traits as personality or musical talent. Geneticists seek to understand how the information encoded in genes is used and controlled by cells and how it is transmitted from one generation to the next. Geneticists also study how tiny variations in genes can disrupt an organism's development or cause disease. Increasingly, modern genetics involves genetic engineering, a technique used by scientists to manipulate genes.

Genetic engineering has produced many advances in medicine and industry, but the potential for abuse of this technique has also presented society with many ethical and legal controversies. Genetic information is encoded and transmitted from generation to generation in deoxyribonucleic acid (DNA). DNA is a coiled molecule organized into structures called chromosomes within cells. Segments along the length of a DNA molecule form genes.

Genes direct the synthesis of proteins, the molecular laborers that carry out all life-supporting activities in the cell. Although all humans share the same set of genes, individuals can inherit different forms of a given gene, making each person genetically unique.

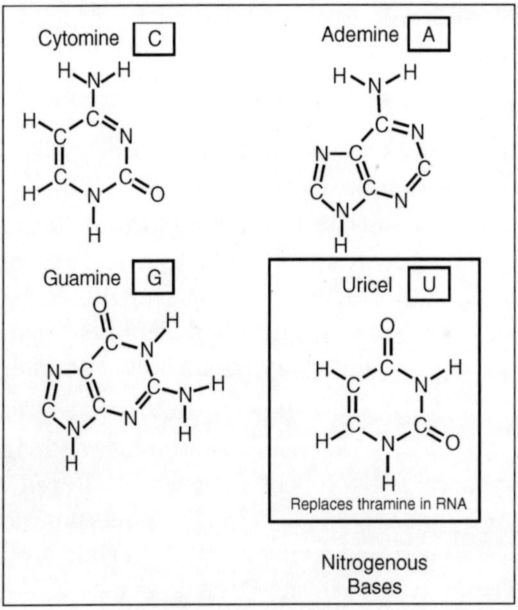

Figure: *Genetics*

Since the earliest days of plant and animal domestication, around 10,000 years ago, humans have understood that characteristic traits of parents could be transmitted to their offspring. The first to speculate about how this process worked were Greek scholars around the 4th century BC, who promoted theories based on conjecture or superstition. Some of these theories remained in favour for several centuries. The scientific study of genetics did not begin until the late 19th century. In experiments with garden peas, Austrian monk Gregor Mendel described th patterns of inheritance, observing that traits were inherited as separate units. These units are now known as genes. Mendel's work formed the foundation for later scientific achievements that heralded the era of modern genetics.

Principles of Heredity

The site where genes work is the cell. Some organisms, such as paramecia or amoebas, are made up of a single cell. Other organisms are made of many kinds of cells, each having a different function. For instance, a tree contains some cells that form the root system and other cells that form leaves. Each cell's function within an organism is determined by the genetic information encoded in DNA.

In animals, plants, and othe eukaryotes (organisms whose cells contain a nucleus), DNA resides within membrane-bound structures in the cell. These structures include the nucleus, the energy-producing mitochondria, and, in plants, the chloroplasts (structures where photosynthesis takes place). In prokaryotes, one-celled organisms and bacteria that lack internal membrane-bound structures, DNA floats freely within the cell body.

The Importance of Genetics

The modern science of genetics influences many aspects of daily life, from the food we eat to how we identify criminals or treat diseases. In agriculture, genetic advances enable scientists to alter a plant or animal to make it more useful. For instance, some food crops, such as oranges, potatoes, wheat, and rice, have been genetically altered to withstand insect pests, resulting in a higher crop yield. Tomatoes and apples have been modified so that they resist discoloration or bruising on their way to market, enhancing their appeal on supermarket shelves.

The genetic makeup of cows has been modified to increase their milk production, and cattle raised for beef have been altered so that they grow faster. Genetic technologies have also helped convict criminals. DNA recovered from semen, blood, skin cells, or hair found at a crime scene can be analysed in a laboratory and compared with the DNA of a suspect. An individual's DNA is as unique as a set of fingerprints, and a DNA match can be used in a courtroom as evidence connecting a person to a crime. Genetics has revolutionized the way industries produce certain substances, many of which formerly required costly and arduous manufacturing methods.

In medicine, scientists can genetically alter bacteria so that they mass-produce specific proteins, such as insulin used by people with diabetes mellitus or human growth hormone used by children who suffer from growth disorders. In other medical applications, genetic technologies have been instrumental in the development of gene therapy. In this still-experimental form of treatment, scientists try to cure disease by replacing malfunctioning genes with healthy ones. Gene therapy has shown promise in treating some devastating conditions, including some forms of cancer and cystic fibrosis.

Genetically engineered vaccines are being tested for possible use against the human immunodeficiency virus (HIV), the virus that causes acquired immunodeficiency syndrome (AIDS). The field of human genetics has been energized in recent years by the Human Genome Project, an international collaboration of scientists, governments, and drug companies from around the world. Scientists working on this project have developed detailed maps that identify the chromosomal locations of the estimated 20,000 to 25,000 human genes. The vast databases emerging from the project help scientists study previously unknown genes as well as many genes all at once to examine how gene activity can cause disease. Scientists expect that the project will lead to the development of new drugs targeted to specific genetic disorders. Despite the benefits derived from genetic advancements, some observers have voiced concerns that genetically engineered organisms could harm people or the environment. Others fear that new genetic technologies may enable scientists to modify genes that affect characteristics other than those responsible for disease. They warn that determining who has undesirable genetic characteristics may lead to discriminatory practices.

Others are concerned about the common misperception that a person's genes determine all aspects of a person's life, including health and behaviour. This misperception leads people to blame their genetic makeup for problems, leaving no room for the influence of free will, personal responsibility, or hope for change. These and other challenging issues place geneticists at the crossroads of science and social responsibility, where they work to promote understanding of genetic advances and prevent the abuse of them.

Cell Division and Reproduction

Organisms could not grow or function properly if the genetic information encoded in DNA was not passed from cell to cell. DNA is packaged into structures called chromosomes within a cell. Every chromosome in a cell contains many genes, and each gene is located at a particular site, or locus, on the chromosome. Chromosomes vary in size and shape and usually occur in matched pairs called homologues. The number of homologous chromosomes in a cell depends upon the organism—for example, most cells in the human body contain 23 pairs of chromosomes, while most cells of the fruit fl *Drosophila* contain 4 pairs. Within all organisms, cells divide to produce new cells, each of which requires the genetic information found in DNA.

Yet simply splitting the DNA of a dividing cell between two new cells would lead to disaster—the two new cells would have different instructions and each subsequent generation of cells would have less and less genetic information to work with. Imagine how chaotic it would be to rip an architectural blueprint in two, give each half to different contractors, and tell them to construct identical buildings. Just as each contractor would require a full copy of the blueprint to construct a complete building, each new cell needs a complete copy of an organism's genetic information to function properly. Organisms use two types of cell division to ensure that DNA is passed down from cell to cell during reproduction.

Simple one-celled organisms and other organisms that reproduce asexually—that is, without the joining of cells from two different organisms—reproduce by a process called mitosis. During mitosis a cell doubles its DNA before dividing into two cells and distributing the DNA evenly to each resulting cell. Organisms that reproduce sexually use a different type of cell division. These organisms produce special cells called gametes, or egg and sperm. In the cell division known as meiosis, the chromosomes in a gamete cell are reduced by half. During sexual reproduction, an egg and sperm unite to form a zygote, in which the full number of chromosomes is restored.

Mitosis

Mitosis occurs in five stages: prophase, prometaphase, metaphase, anaphase, and telophase. During prophase, the start of mitosis, the DNA of each chromosome replicates. Each chromosome then reorganizes into paired structures called sister chromatids, with each member of the pair containing a full copy of the DNA sequence. The sister chromatids condense, thickening until they appear joined at a single site, known as the centromere. Prometaphase is marked by the disintegration of the nuclear membrane. The sister chromatids line up in the middle of the cell halfway between the poles. In metaphase, exactly half of the chromatids face one pole, and half face the other pole. This equilibrium position is called the metaphase plate. In anaphase, the chromatid pairs split apart at the centromere, and each half of the pair then moves towards opposite poles of the cells. In telophase, the final stage of mitosis, a nuclear membrane forms around the chromosomes at each pole of the cell. Mitosis ends with the formation of two new cells, each with a matching full set of chromosomes as well as an identical complement of cellular structures.

Meiosis

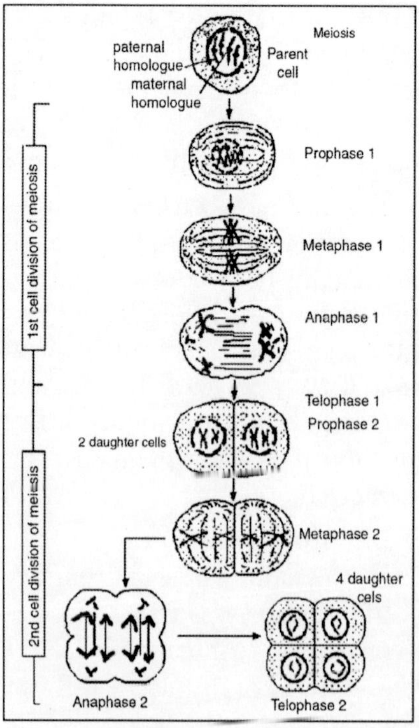

Figure: *Meiosis*

During meiosis, two cell divisions occur to produce four daughter cells from the original parent cell. Each resulting cell has half the chromosomal DNA of the parent cell. A half set of chromosomes in an organism is known as the haploid number. In the first cell division of meiosis the chromosomes of a gamete cell duplicate and join in pairs. The paired chromosomes align at the equator of the cell, and then separate and move to opposite poles in the cell. The cell then splits to form two daughter cells. As meiosis proceeds, the two daughter cells undergo another cell division to form four cells, each of which bears half of the number of chromosomes found in the other cells of the organism. Meiosis ensures that reproduction will produce a zygote that has received one set of chromosomes from the male parent and one set of chromosomes from the female parent to form a full set of chromosomes. The entire set of chromosomes in an organism is known as the diploid number. Once formed, the zygote continues to divide and grow through the process of mitosis

Patterns of Inheritance

In life forms that reproduce asexually, such as bacteria and amoebas, all offspring share the exact same genes and are identical to their parents. The genetic transmission that occurs in organisms that reproduce sexually is far more complex. An individual that forms by the union of two gametes inherits its chromosomes from two distinct parents.

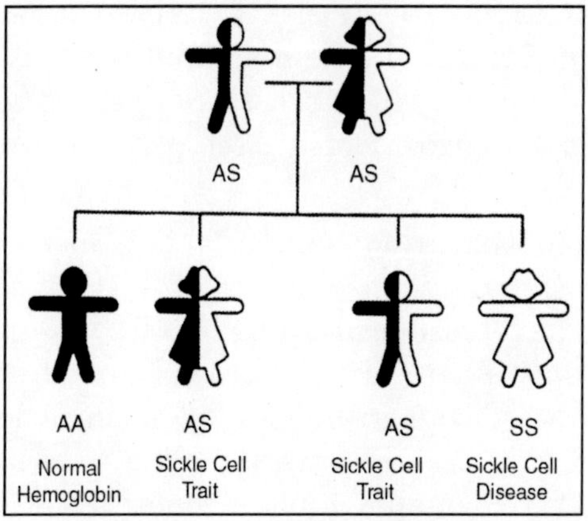

Figure: *Patterns of Inheritance*

Consequently, sexual reproduction guarantees that offspring with new combinations of genes will continually arise. Certai patterns of inheritance were evident long before scientists discovered the molecular structure of DNA and chromosomes. Throughout history, people have recognized that certain traits, whether in humans, animals, or agricultural crops, could be passed from generation to generation. Yet for centuries, people were unable to reconcile many confusing observations about the mechanisms of inheritance. The first person to make sense of this complex subject was Austrian monk Gregor

Mendel, who conducted a series of experiments on pea plants beginning in the 1850s. Mendel observed the results of crossbreeding plants with different characteristics, such as height, flower colour, and seed shape. His conclusions from these experiments led him to develop explanations for how traits are transmitted from generation to generation. Mendel's theories form the foundation of modern genetics.

Mendel's Rules

In his research, Mendel observed that characteristics were inherited as separate units, each of which was inherited independently of the others. Mendel suggested that each parent has pairs of these units but contributes only one of each pair to offspring. The units that Mendel described were later given the name *genes.* Mendel recognized that a gene can exist in different forms. Today these alternate forms are known as alleles. For example, pea seeds, the edible part of the plant we call peas, have a texture trait controlled by a single gene.

This gene occurs in two alleles: one corresponding to round (smooth) peas, the other to wrinkled peas. Although an individual can carry only two alleles for a particular gene, each gene may have dozens of different alleles. Mendel's experiments focused on interbreeding different strains of pea plants and then observing the traits that appeared in subsequent generations. When he crossbred plants with round peas and those with wrinkled peas, he discovered that all of the resulting offspring had round peas. Today we know that peas are round and smooth when they contain the right amount of sugar. If peas are missing the gene that produces a protein called starch branching enzyme 1 (SBE1), the peas make too much sugar, causing the peas to swell and then wrinkle and shrivel as they dry.

Mendel concluded that when an organism has two different alleles corresponding to the same genetic trait, one of the two may be dominant. The other allele is said to be recessive, meaning that its presence will be detectable only if an organism has inherited the recessive gene from both parents. For convenience, geneticists designate alleles by a single letter—the dominant allele is represented by a capital letter and the recessive allele by a small letter. In the pea texture example, a plant inherits one allele for pea texture from each parent. The dominant allele that produces SBE1, resulting in ound, smooth peas, is designated as *R*, while the recessive allele that does not produce SBE1 and produces wrinkled peas is designated as *r*. To determine the set of alleles an organism has for a given trait just by visual observation can often be difficult. In the pea plant example, for instance, plants with smooth peas might be carrying two dominant alleles for that characteristic (RR) or one dominant and one recessive allele (Rr). Geneticists use the term genotype to refer to the combination of genes that code for a trait, while the termphenotype describes the physical manifestation of that trait.

Therefore, the presence of two dominant alleles for pea texture (*RR*) would reflect the genotype while a smooth pea indicates thephenotype. Mendel did not limit his experiments to testing the rules of inheritance of single traits. He also studied plant traits involving multiple pairs of genes, breeding plants that have round, yellow seeds with plants that produce wrinkled, green seeds. Such experiments demonstrated that thepatterns of inheritance he observed in his experiments with single traits also apply to cases involving more complex gene combinations.Exceptions to Mendel's Rules

Mendel published his studies in a science journal in 1865, at which time no other scientist commented on his work. Since that time, geneticists have learned that sometimes genes do not easily conform to so-called Mendelianpatterns of inheritance.

Incomplete Dominance

In cases of incomplete dominance, the inheritance of a dominant and a recessive allele results in a blending of traits to produce intermediate characteristics. For example, four-o'clock paint plants may have red, white, or pink flowers. Plants with red flowers have two copies of the dominant allele *R* for red flower colour (*RR*). Plants with white flowers have two copies of the recessive allele *r* for white flower colour (*rr*). Pink flowers result in plants with one copy of each allele (*Rr*), with each allele contributing to a blending of colors.Quantitative Inheritance

Mendel focused his studies on traits determined by a single pair of genes, and the resultingphenotype was easy to distinguish. A tall plant can be markedly different from a short one, and a green pea can easily be distinguished from a yellow one. There are some traits, however, that are not easy to distinguish. Human skin colour, for example, may be any of a wide variety of shades. Traits such as skin colour differ from the ones Mendel studied because they are determined by more than one pair of genes. In this form of inheritance, known asquantitative inheritance, each pair of genes has only a slight effect on the trait, while the cumulative effect of all the genes determines the physical characteristics of the trait. At least four pairs of genes control human skin colour. Multiple genes also control many traits important in agriculture, such as milk production in cows and ear length in corn.

Multiple Alleles

Another exception to Mendelian genetics involves genes with multiple alleles. Certain traits are controlled by multiple alleles that have complex rules of dominance. In humans, for example, the gene for blood type has three alleles: I_A, I_B, and i. With three alternatives for each member of a gene pair, there are six possible combinations of these genes (I_AI_A, I_BI_B, ii, I_Ai, I_Bi, I_AI_B). Although there are six possible combinations, humans have only four major blood types: A, B, AB, and O. This results because both I_A and I_B dominate over i, but not over each other, so a person with a gene combination of I_AI_A or I_Ai has blood type A. The gene combinations I_BI_B and I_Bi both produce blood type B. I_AI_B results in a blood type AB, and ii results in blood type O.

Gene Linkage

In his experiments, Mendel was careful to study traits in pea plants where one trait did not appear to influence another, such as the plant's height or the pea's texture. These twophenotypes (height and texture) occur randomly with respect to one another in a manner known as independent assortment. Today scientists understand that independent assortment occurs when the genes affecting thephenotypes are found on different chromosomes.

An exception to independent assortment develops when genes appear near one another on the same chromosome. When genes occur on the same chromosome, they are inherited as a single unit. Genes inherited in this way are said to be linked. For example, in fruit flies the genes affecting eye colour and wing length are inherited together because they appear on the same chromosome. But in many cases, genes on the same chromosome that are inherited together produce offspring with unexpected allele combinations. This results from a process calledcrossing over. Sometimes at the beginning of meiosis, a chromosome pair (made up of a chromosome from the mother and a chromosome from the father) may intertwine and exchange sections of chromosome.

The pair then breaks apart to form two chromosomes with a new combination of genes that differs from the combination supplied by the parents. Through this process of recombining genes, organisms can produce offspring with new combinations of maternal and paternal traits that may contribute to or enhance survival.

Sex-Linked Traits

Most chromosome pairs consist of identical, or homologous, partners. In many species, including humans, there is one pair of chromosomes in which the partners noticeably differ from each other. These are called the sex chromosomes because they determine the differences between males and females.

		♀	
		×	$\underline{\times}$
♂	$\underline{\times}$	$\underline{\times}$ ×	$\underline{\times\times}$

Figure: *Sex-Linked Traits*

Genes located on the sex chromosomes display differentpatterns of inheritance than genes located on other chromosomes. In human females, the sex chromosomes consist of two X chromosomes, while males have an X chromosome and a shorter Y chromosome with many fewer genes. In males the X chromosome contains many genes that have no corresponding gene on the Y chromosome. A male's X chromosome may contain a recessive allele associated with a genetic disorder, such as hemophilia or Duchenne muscular dystrophy. In this case, males do not have a normal second copy of the gene on the Y chromosome to mask the effects of the recessive gene, and disease typically results. Additional examples of sex-linked traits include red-green colour blindness in humans and eye colour in fruit flies.

The Genetic Code

The structure of DNA encodes all the information every cell needs to function and thrive. In addition, DNA carries hereditary information in a form that can be copied and passed intact from generation to generation. A gene is a segment of DNA. The biochemical instructions found within most genes, known as the genetic code, specify the chemical structure of a particular protein. Proteins are composed of long chains of amino acids, and the specific sequence of these amino acids dictates the function of each protein.

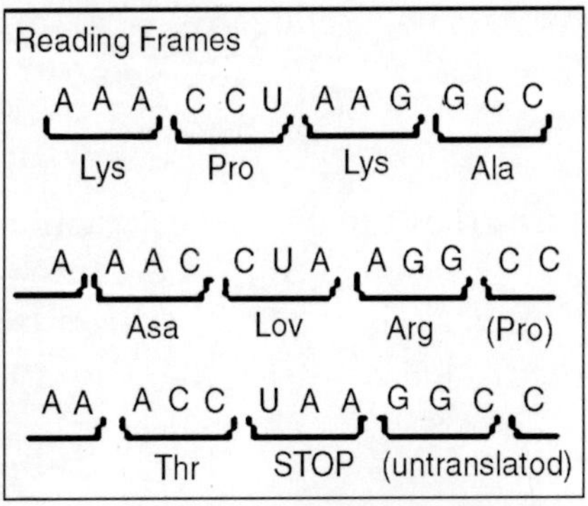

Figure: *Genetic Code*

The DNA structure of a gene determines the arrangement of amino acids in a protein, ultimately determining the type and function of the protein manufactured.

DNA Structure

DNA molecules form from chains of building blocks called nucleotides. Each nucleotide consists of a sugar molecule called deoxyribose that bonds to a phosphate molecule and to a nitrogen-containing compound, known as a base. DNA uses four bases in its structure: adenine (A), cytosine (C), guanine (G), and thymine (T). The order of the bases in a DNA molecule—the genetic code—determines the amino acid sequence of a protein. In the cells of most organisms, two long strands of DNA join in a single molecule that resembles a spiraling ladder, commonly called a double helix. Alternating phosphate and sugar molecules form each side of this ladder. Bases from one DNA strand join with bases from another strand to form the rungs of the ladder, holding the double helix together. The pairing of bases in the DNA double helix is highly specific—adenine always joins with thymine, and guanine always links to cytosine. These base combinations, known as complementary base pairing, play a fundamental role in DNA's function by aiding in th replication and storage of genetic information.

Complementary base pairing also enables scientists to predict the sequence of bases on one strand of a DNA molecule if they know the order on the corresponding, or complementary, DNA strand. Scientists use complementary base pairing to help identify the genes on a particular chromosome and to develop methods used in genetic engineering. Genes line up in a row along the length of a DNA molecule. In humans a single gene can vary in length from 100 to over 1,000,000 bases. Genes make up less than 2 per cent of the length of a DNA molecule. The rest of the DNA molecule is made up of long, highly repetitive nucleotide sequences. Once dismissed as "junk" DNA, scientists now believe these nucleotide sequences may play a role in the survival of cells. Identifying the function of these sequences is a thriving field of genetics research

Protein Synthesis

DN replication ensures that the genetic instructions encoded in DNA can be used continuously through generations to produce the proteins that build and operate the cells of an organism. The process of tapping the genetic code to create proteins, known a protein synthesis, has two crucial steps: transcription and translation.

DN Replication

In order for inherited traits to be transmitted from parent to child, the genetic information encoded in DNA must be copied with great precision during cell division. The accuracy of DN replication depends upon the complementary pairing of bases. Durin replication, the DNA double helix unwinds and bonds joining the base pairs break, separating the DNA molecule into two separate strands. Each strand of DNA directs the synthesis of another complementary strand. The unpaired bases of each DNA strand attach to bases floating within the cell. But the DNA strand's unpaired bases bond only with specific, complementary bases—for example, an adenine base will bond only with a thymine base and a cytosine bases will pair only with a guanine base. Once all of the bases of a DNA strand bond to complementary bases, the complementary bases then link to each other, forming a new DNA double-helix molecule. Thus the original DNA molecule replicates into two DNA molecules that are exact duplicates.

Transcription

Transcription transfers the genetic code from a molecule of DNA to an intermediary molecule called ribonucleic acid (RNA). The basic nucleotide structure of RNA resembles that of DNA, but the two compounds have three critical differences. First, the structure of RNA incorporates the sugar ribose rather than deoxyribose, the sugar in DNA. Second, RNA uses the base uracil (U) instead of thymine (T). In RNA uracil binds with adenine just as thymine does in DNA. Third, RNA usually exists as a single strand, unlike the double-helix structure that normally characterizes DNA. Transcription involves the production of a special kind of RNA known as messenger RNA (mRNA).

The process begins when the two strands of a DNA molecule separate, a task directed by the enzyme RNA polymerase. After the double helix splits apart, one of the strands serves as a template, or pattern, for the formation of a complementary mRNA molecule. Free-floating individual bases within the cell bind to the bases on the DNA template using complementary base pairing. The individual bases then link together to form a strand of mRNA. I eukaryotes (organisms whose cells have a nucleus), the mRNA strand undergoes an additional step before the next stage o protein synthesis can occur. The mRNA strand consists of coding regions called exons separated by regions called introns. The introns do not contribute t protein synthesis. Special enzymes in the nucleus remove the introns from the mRNA strand. The remaining exons then link together to form an mRNA strand that contains the entire code for making a protein.

Translation

Once transcription is complete and the genetic code has been copied onto mRNA, the genetic code must be converted into the language of proteins. That is, the information coded in the four bases found in mRNA must be translated into the instructions encoded by the 20 amino acids used in the formation of proteins. This process, called translation, takes place in cellular organelles called ribosomes. I eukaryotes, mRNA travels out of the nucleus into the cell body to attach to a ribosome. In prokaryotes (organisms without a nucleus), the ribosome clasps mRNA and starts translation before these strands have finished transcription and separated from the DNA. In bot eukaryotes and prokaryotes, the ribosome acts like a workbench and clamp that holds the mRNA strand and coordinates the activity of enzymes and other molecules essential to translation. Another form of RNA called transfer RNA (tRNA) is found in the cytoplasm of the cell. There are many different types of tRNA, and each type binds with one of the 20 amino acids used in protein formation. One end of a tRNA binds with a specific amino acid. The other end carries three bases, known as an anticodon. The tRNA with an amino acid attached travels to the ribosome where the mRNA is stationed.

The anticodon of the tRNA undergoes complementary base pairing with a series of three bases on the mRNA, known as the codon. The mRNA codon codes for the type of amino acid carried by the tRNA. A second tRNA bonds with the next codon on the mRNA. The resident tRNA transfers its amino acid to the amino acid of the incoming tRNA and then leaves the ribosome.

This process continues repeatedly, with new tRNA receiving the growing chain of amino acids, known as a polypeptide chain, from a resident tRNA. The ribosome moves the mRNA strand one codon at a time, making new codons available to bind with tRNAs. The process ends when the entire sequence of mRNA has been translated. The polypeptide chain falls away from the ribosome as a newly formed protein, ready to go to work in the cell.

Mutations

Occasionally mistakes occur during DN replication an protein synthesis. Any alteration in the structure of a gene results in a mutation. Mutations occur during DN replication when the chemical structure of genes undergoes random modifications. Once a change has occurred, the altered genes continue to replicate in their changed form unless another mutation occurs. Sometimes mutations occur during transcription or translation, causin protein synthesis to go awry. Although mutations may occur in any living cell, they are most important when they occur in gametes because then the change affects the traits of following generations.

Most mutations harm an organism. If a mutation occurs in a gene sequence that codes for a particular protein, the mutation may result in a change in the amino acid sequence directed by the gene. This change, in turn, may affect the function of the protein. The implications can be significant: The amino acid sequence distinguishing normal hemoglobin from the altered form of hemoglobin responsible for sickle-cell anemia differs by only a single amino acid. Some mutations may be neutral or silent and do not affect the function of a protein. Occasionally a mutation benefits an organism. Over the course of evolutionary time, however, mutations serve the crucial role of providing organisms with previously nonexistent proteins.

In this way, mutations are a driving force behind genetic diversity and the rise of new or more competitive species better able to adapt to changes, such as climate variations, depletion of food sources, or the emergence of new types of disease. Mutations can produce a change in any region of a DNA molecule. In *point mutation*, for example, a single nucleotide replaces another nucleotide. Although point mutation produces a small change to the DNA sequence, it may cause a change in the amino acid sequence, and thus the function, of a protein. Far more serious are mutations that involve the addition or deletion of one or more bases from a DNA molecule.

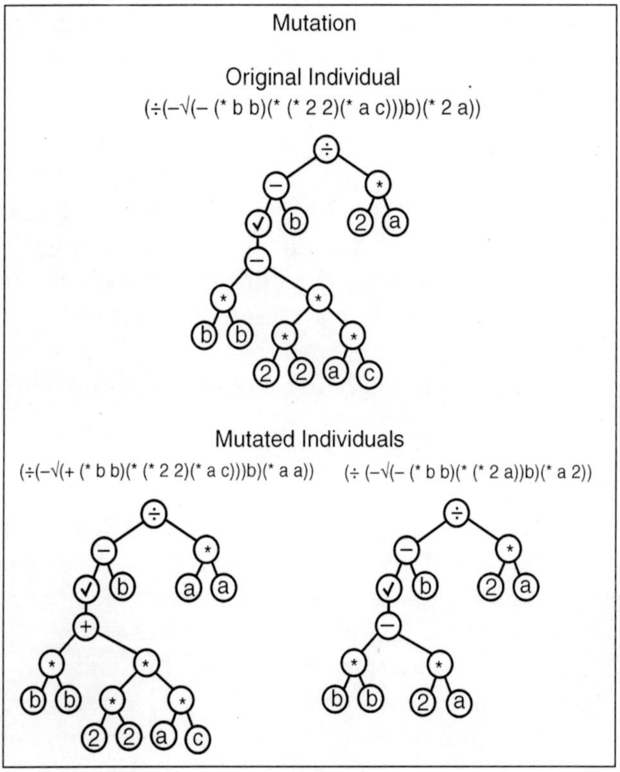

Figure: *Mutations*

Adding or subtracting even a single base from a normal sequence during transcription can disrupt translation by shifting the "reading frame" of every subsequent codon. For example, an mRNA strand may include two codons in the following sequence: AUG UGA. The addition of a cytosine base at the beginning of this sequence shifts the "spelling" of these codons so that they read: CAU GUG.

This may result in an incorrect amino acid sequence during translation, or the protein may be truncated. Known a frameshift mutation, this type of alteration could result in the production of a protein with no real function or one with a harmful effect. Sometimes mutations are caused b transposition, in which long stretches of DNA (containing one or more genes) move from one chromosome to another. These jumping genes, calle transposons, can disrupt transcription and change the type of amino acids inserted into a protein Transposons rearrange and interrupt genes in a way that generally improves the genetic variation of a species. While mutations can occur spontaneously, some can be caused by exposure to physical or chemical agents in the environment called mutagens. Common environmental mutagens include ultraviolet rays from the sun and various chemicals, such as asbestos, cigarette smoke, and nitrous acid. High-energy radiation, such as medical X rays, can cause DNA strands to break, leading to the deletion of potentially important genetic information.

Radiation damage can also affect an entire chromosome, disrupting the function of many genes. In chromosomal translocation, a piece of one chromosome breaks off and merges with another chromosome. In some cases, large sections of chromosomes may break off and be lost. The cell has highly effective self-repair mechanisms that can correct the harmful changes made by mutations and prevent some mutations from being passed on. Some 50 specialized enzymes locate different types of faulty sequences in the DNA and clip out those flaws. Another repair mechanism scans DNA afte replication and marks mismatched base pairs for repair.

Gene Regulation

The processes that enable information to be copied from genes and then used to synthesize proteins must be regulated if an organism is to survive. Different cells within an organism share the same set of chromosomes. In each cell some genes are active while others are not. For example, in humans only red blood cells manufacture the protein hemoglobin and only pancreas cells make the digestive enzyme known as trypsin, even though both types of cells contain the genes to produce both hemoglobin and trypsin. Each cell produces different proteins according to its needs so that it does not waste energy by producing proteins that will not be used. A variety of mechanisms regulate gene activity in cells. One method involves turning on or off gene transcription, sometimes by blocking the action of RNA polymerase, an enzyme that initiates transcription. Gene regulation may also involve mechanisms that slow or speed the rate of transcription, using specialized regulatory proteins that bind to DNA. Depending on an organism's particular needs, one regulatory protein may spur transcription for a particular protein, and later, another regulatory protein may slow or halt transcription.

Prokaryotes

The bacteriu *Escherichia coli* (commonly referred to as *E. coli*), found in the intestines of humans and other mammals, provides a good example of gene regulation. *E. coli* uses three enzymes to digest lactose, the primary sugar found in milk. The bacterium produces great quantities of lactose-digesting enzymes when lactose is present and saves energy by not synthesizing the enzymes when the sugar is not available. *E. coli* prefers consuming glucose to lactose, so the bacterium produces these enzymes in the presence of lactose only when no glucose is available. A region of DNA known as an operon controls this gene regulation process. In *E. coli*, the operon includes at least five genes: Three genes, called lac genes, code for the enzymes that digest lactose; one gene encodes for a regulatory protein, called a repressor, that can sense the presence or absence of lactose; and one gene, called an operator, activates transcription, the first step in the synthesis of the lactose-digesting enzymes.

In the absence of lactose, the repressor protein binds with the operator gene to block transcription. This prevents the lac genes from being transcribed and halts production of the lactose-digesting enzymes. If lactose is present, the repressor protein binds to the sugar, leaving the operator gene free to trigger transcription of the lac genes. The transcription of the lac genes produces the mRNA that will direct the production of the three lactose-digesting enzymes

Eukaryotes

Gene regulation i eukaryotes is more complex than in bacteria and other prokaryotes. Even the simples eukaryotes have far more genes than prokaryotes, and these genes must be turned on or off as conditions dictate. Most multicellular organisms contain different types of cells that serve specialized functions. The cells of an animal's heart, blood, skin, liver, and muscles all contain the same genes. But in order to carry out their specific functions within the body, each cell must produce different proteins and respond to changing environmental stimuli, such as glucose levels in the blood or body temperature.

Such specialization is possible only with sophisticated gene regulation Eukaryotes use a variety of mechanisms to ensure that each cell uses the exact proteins it needs at any given moment. In one method, eukaryotic cells use DNA sequences called enhancers to stimulate the transcription of genes located far away from the point on the chromosome where transcription occurs. If a specific protein binds to an enhancer site on the DNA, it causes the DNA to fold so that the enhancer site is brought closer to the site where transcription occurs.

This action can activate or speed up transcription in the genes surrounding the enhancer site, thereby affecting the type and quantity of proteins the cell will produce. Enhancers often exert their effects on large groups of related genes, such as the genes that produce the set of proteins that form a muscle cell. Gene regulation can also take place after transcription has occurred by interfering xwith the steps that modify mRNA before it leaves the nucleus to take part in translation. This process typically involves removing exons (segments that code for specific proteins) and introns.

These sections of the mRNA can be modified in more than one way, enabling a cell to synthesize different proteins depending on its needs.

Genes in Development

Gene regulation helps individual cells within an organism function in a specialized way. Other regulatory mechanisms coordinate the genes that determine how cells develop. All of the specialized cells in an organism, including those of the skin, muscle, bone, liver, and brain, derive from identical copies of a single fertilized egg cell. Each of these cells has the exact same DNA as the original cell, even though they have vastly different appearances and functions. Genes dictate how these cells specialize.

Early in an organism's embryonic development the overall body plan forms. Individual cells commit to a particular layer and region of the embryo, often migrating from one location to another to do so. As the organism grows, cells become part of a particular body organ or tissue, such as skin or muscle. Ultimately, most cells become highly specialized—not only to develop into a neuron rather than a muscle cell, for example, but to become a sensory neuron instead of a motor neuron. This process of specialization is called differentiation. At each stage of the differentiation process, specific genes known as developmental control genes actively turn on and switch off the genes that differentiate cells. One class of developmental control genes, known as homeotic genes, directs the formation of particular body parts. Activating one set of homeotic genes instructs part of an embryo to develop into a leg, for example, while another set initiates the formation of the head.

If a homeotic gene becomes altered or damaged, an organism's body development can be dramatically disrupted. A change in a single gene in some insects, for instance, can cause a leg to grow where an antenna belongs. Homeotic genes work by regulating the activity of other genes. Homeotic genes code for the production of a regulatory protein that can bind to DNA and thus affect the transcription of one or more genes. This enables homeotic genes to initiate or halt the development and specialization of characteristics in an organism.

Nearly identical homeotic genes have been identified in varied organisms, such as insects, worms, mice, birds, and humans, where they serve similar embryonic development functions. Scientists theorize that homeotic genes first appeared in a single ancestor common to all these organisms. Sometime in evolutionary history, these organisms diverged from their common ancestor, but the homeotic genes continued to be passed down through generations virtually unchanged during the evolution of these new organisms.

9

An Introduction to Restoration Genetics

Why is Genetic Diversity Important?

All living organisms carry a genetic blueprint. This is so regardless of whether they are plants, animals, or fungi, whether they are short- or long-lived, and whether they reproduce sexually or clonally. Therefore, to the extent that restoration deals with living organisms, genetics are part of the picture. Although the basic principles underlying restoration genetics may be familiar, to date surprisingly little attention has been devoted to genetic considerations in restoration practice. The purpose of this Restoration Science and Policy Paper is to outline some considerations that restoration designers and managers should be aware of, and to identify more detailed resources that may be useful in practice.

Genotypes Partly Determine Organisms' Physical form and Function

Biologists refer to two basic expressions of variation, the genotype and the phenotype. The *genotype* is the genetic code of an organism; in organisms with nucleated cells (most multi-cellular plants, animals, and fungi), the essential code is found in the nucleus of each cell. Additional genetic codes reside in other components of the cell, such as mitochondria (in animals) and chloroplasts (in plants). An organism's genotype consists of a large number of *genes* (50,000-100,000 in a typical vertebrate), which can be at multiple sites (*loci*) on chromosomes. Genes have a variety of functions, the most important of which is to code for the production of specific amino acids, which are ultimately used to synthesize proteins. In most higher plants and animals (but not fungi, bryophytes, and many marine invertebrates), the adult phase has two copies of each gene, one derived from each parent. When the two copies are the same, the individual is *homozygous* for that gene; if the two copies are different, the individual is *heterozygous* for that gene. The various forms of a gene are referred to as *alleles*; when these forms are identical across a population, the gene is considered *monomorphic*, and if more than one allele exists the gene is considered *polymorphic*.

Phenotype is the expression of these genes as a living organism in a particular environment, and is influenced by environmental context at every level from the cell to the whole organism. It is frequently very difficult to separate variation in an organism's traits with a genetic basis from traits that are phenotypically variable; mistakes are Introduction to restoration genetics. Falk, Knapp, and Guerrant, Society for Ecological Restoration, 17 December 2001. Commonly made both ways (assuming a genetic basis when observed variation is really phenotypic plasticity; underestimating subtle genetic effects on traits that are assumed to be simply "phenotype"). For example, a plant growing in a poor environment might end up having a small and stunted phenotype, despite having "good " genes. This does not mean that genetic variation is not important to the fitness of individuals or populations. Both the environment and the genes will ultimately contribute to restoration success, but steps to ensure an optimal genetic makeup of a population are commonly overlooked.

Examples abound in the scientific literature illustrating how genetic composition affects the form and function of organisms. In fact, the recognition of genetic variation among individuals was a primary insight that led to the formulation of evolutionary theory as we know it today (Freeman and Herron 1998). Genes regulate body size, shape, physiological processes, behavioral traits, reproductive characteristics, tolerance of environmental extremes, dispersal and colonizing ability, the timing of seasonal and annual cycles (*phenology*), disease resistance, and many other traits. Thus, to ignore genetic variation in ecology is to ignore one of the fundamental forces that shape the biology of living organisms. From a restoration perspective, organism and population genetics are fundamental considerations in the design, implementation, and expectations of any project, whether or not explicit consideration is given to the genetic dimension.

Genetic Diversity Helps Organisms Cope with Current Environmental Variability

Organisms exist in environments that vary in time and over space. Such variation is often described in terms of the *natural* or *historic range of variability* (NRV, HRV) in environmental conditions such as weather, disturbance events, resource availability, population sizes of competitors, etc. If a group of organisms (say, a population of species X) were to live in a completely stable physical and biological environment, then a relatively narrow range of phenotypes might be optimally adapted to those conditions. Under these circumstances, Species X would benefit more by maintaining a narrow range of genotypes adapted to prevailing conditions, and allele frequencies might eventually attain equilibrium. By contrast, if the environment is Introduction to restoration genetics.

Examples of traits with a genetic basis for tolerance of environmental variation important in restoration work include tolerance of freezing, drought or inundation, high or low light availability, salinity, heavy metals, soil nutrient deficiencies, and extreme soil pH values in plants; resilience to fluctuating temperature, dissolved oxygen, and nutrient availability in aquatic organisms; and resistance to novel diseases in all groups of organisms (Huenneke 1991). For example, if all individuals in a population are the same genotype with limited drought tolerance, then a single climatic event may destroy the entire population. Plant populations often include individuals with a range of flowering or emergence times. For instance, Great Basin shrub populations include individuals that leaf out and flower over a period of weeks, increasing the likelihood of persistence of the population through periods of unusually early or late growing conditions. Knapp *et al.* (2001) documented flowering periods in a population of individual blue oak trees and found that trees initiated flowering over a period of a month in the spring. Such variability could potentially be adaptive, since it is more likely that at least some trees in the population will flower during warm sunny periods when wind pollination is most successful.

A diverse array of genotypes appears to be especially important in disease resistance. Genetically uniform populations (such as highly inbred crops) are famously vulnerable to diseases and pathogens, which can (and do) decimate populations in which all individuals are equally vulnerable. Such uniformity also predisposes a population to transmit disease from one individual to another: instead of having isolated diseased individuals, nearly every individual may be exposed to disease by direct contact or proximity. More diverse populations are more Introduction to restoration genetics. Falk, Knapp, and Guerrant, Society for Ecological Restoration, 17 December 2001, likely to include individuals resistant to specific diseases; moreover, infected individuals occur at lower density, and thus diseases or pathogens may move more slowly through the population.

Finally, genetic variation is a factor in competition among individuals in real ecological communities. Traits with a genetic basis such as flower size are key factors in competition among individuals. Among animals, behavioral traits may regulate interspecific competition. Since organisms make energetic or life history tradeoffs among traits (for example, allocating energy between growth and reproduction), genetic variability is an important factor in how populations function.

Diversity within Populations Reduces Potentially Deleterious Effects of Breeding Among Close Relatives

In addition to its adaptive value at the population level, genetic variation (or its lack) within individuals can affect their survival and performance. When both copies of a gene (in a diploid organism) are identical (*i.e.*, when an individual is *homozygous* at that gene or locus), the expression of that gene may include traits that are less beneficial to survival or reproduction in particular circumstances. This may lead to physiological or behavioral problems of genetic origin, such as malformed physical structure, poor biochemical balance, improper organ formation and function, altered social behaviour, and susceptibility to disease (Chai 1976).

Homozygosity at key gene loci is a common result of *inbreeding*, which is sexual reproduction among closely related individuals. Small populations and species that do not disperse well (or are constrained by a fragmented landscape from exchanging genes with other populations) can be particularly susceptible to *inbreeding depression*, which is reduced overall survival and reproduction of organisms with low heterozygosity. Inbreeding depression arises from a variety of causes, including expression of unfavorable or deleterious alleles, and often leads to lower survival and birth or Introduction to restoration genetics. Falk, Knapp, and Guerrant, Society for Ecological Restoration, 17 December 2001, reproductive rates.

Heterozygosity is not always beneficial, nor does inbreeding always have adverse effects. In some circumstances, a population may be so well adapted to its local circumstances that introducing alleles from other populations actually reduces its performance (*outbreeding depression*) (Templeton 1991). Paradoxically, organisms can experience outbreeding depression for some traits, while suffering the effects of inbreeding depression for other characters. Many organisms (particularly plants) have evolved breeding strategies, such as self-pollination, that allow successful persistence in small populations without apparent short-term effects of inbreeding depression.

Genetic Diversity is the Primary Basis for Adaptation to Future Environmental Uncertainty

Finally, genetic variation holds the key to the ability of populations and species to persist over evolutionary time through changing environments (Freeman and Herron 1998). No organism can predict the future (and evolutionary theory does not require them to), nor can any organism be optimally adapted for all environmental conditions. Nonetheless, the current genetic composition of a species influences how well its members will adapt to future physical and biotic environments.

Consider, for example, the response of a population of plant Species A to a period of rapid climate change. Changes in climate are reflected locally in altered annual rainfall and its seasonal distribution, changes in annual mean maximum and minimum temperatures and their seasonality, changes in prevailing wind direction and speed, and many other factors. These changes cause the zone of suitable climate for Species A to shift geographically. Populations with a diverse array of optima for these climatic conditions (*i.e.*, some individuals that are slightly more cold or heat tolerant, etc.) are more likely to persist, because some individuals will survive the new conditions and reproduce.

If suitable climate zones shift across the landscape, seeds of some individuals may be dispersed into the new location and find good conditions for growing. In this manner, the population can "migrate" across the landscape over generations. By contrast, populations that have a narrower range of genotypes (and are more phenotypically uniform) may simply fail to survive and reproduce at all as conditions become less Introduction to restoration genetics. Falk, Knapp, and Guerrant, Society for Ecological Restoration, 17 December 2001, locally favourable. Such populations are more likely to become *extirpated* (locally extinct), and in extreme cases the entire species may end up at risk of extinction. For example, the Florida Yew (*Torreya taxifolia*) is currently one of the rarest conifer species in North America. But in the early Holocene (10,000 years ago), when conditions in southeastern North America were cooler and wetter than today, the species was probably widespread. For reasons that are not completely understood, *T. taxifolia* failed to migrate northward as climate changed during the Holocene. Today, it is restricted to a few locations in the Apalachicola River Basin in southern Georgia and the Florida panhandle.

As the *T. taxifolia* story illustrates, once species are pushed into marginal habitat at the limitations of their physiological tolerance, they may enter an "extinction vortex", a downward cycle of small populations, reduced genetic variability, reduced ability to adapt to novel conditions, leading to further reductions in population size, and so on. Reduced genetic variability is a key step in the extinction vortex.

How is Genetic Diversity Distributed in Natural Populations?

There are genetic differences among individuals within most (but not all) populations of plants and animals. There are also differences among populations across the range of each species. In this section we review some basic patterns of how genetic diversity of species is distributed, or *partitioned*.

Monomorphic and Polymorphic Alleles

By definition, individuals within a given species share some percentage of alleles; otherwise, they would not be considered members of the same species. This shared (or common) portion of the gene pool includes two basic classes of genes. Genes that are monomorphic (*i.e.*, $A = 1$, where A is the number of alleles per locus) within a species are common to all populations, and indeed essentially to all individuals. (Technically, "monomorphic" genes are defined as those for which the frequency of the most common allele is > 99%; thus, some variation may be present even at these loci). The monomorphic proportion of the total genome varies among taxonomic groups, typically in the range of 85% in animals and 50% in plants. The remainder of loci (on average, 15% in animals and 50% in plants) is Introduction to restoration genetics. Falk, Knapp, and Guerrant, Society for Ecological Restoration, 17 December 2001. polymorphic ($A > 1$), varying among individuals in a population, and among populations within a species. These loci are of most concern to restoration, since polymorphic loci may or may not be adequately represented depending on the design and execution of a restoration program.

Diversity within and Among Populations of a Single Species

The genetic profile of whole populations typically varies from place to place across a species range. These differences may arise as the result of chance occurrences, such as the genetic composition of dispersing individuals that create a new population (*founder effect*), or changes in allele frequencies that result from chance matings in very small populations (*genetic drift*). Differences among populations can also arise systematically, especially if the environment in various places exposes individuals to different optima for survival and reproduction (*fitness*).

For these and other reasons, populations often diverge from one another in their genetic composition. Such divergence is especially strong and rapid when there is little *gene flow* between populations (*e.g.*, limited dispersal of seeds or pollen, or limited movement of animals across physiographic barriers). Over evolutionary time, such among-population genetic differences can accumulate and eventually result in the development of a new species (*allopatric speciation*). Indeed, "populations" are defined as much (or more) by patterns of mating and gene flow as by the physical distribution of individuals, although the two are often closely related (Hartl and Clark 1997).

Each species distributes its genetic diversity (one measure of which is the total of all alleles at all loci) in a pattern reflecting both its biology and its history. For example, nearby populations of plants that are pollinated by bees may share many alleles because genes (packaged in pollen grains) can flow easily between sites. Such species may have fewer unique alleles in each population, so populations tend to be genetically similar. By contrast, there may be less gene flow among populations of species that are pollinated by ground-dwelling flightless beetles, or whose heavy fruits fall to the ground in the vicinity of the parent tree. Gene flow can also be obstructed by physical barriers (*i.e.*, topography or habitat that a pollinator, disperser, or migrating individual cannot cross), as well as by disturbance. Introduction to restoration genetics. Falk, Knapp, and Guerrant, Society for Ecological Restoration, 17 December 2001.

Even if variation within a population is low, there may be considerable variability *among* populations. Imagine Species X, a plant that lives in high mountaintop alpine areas, in which every individual on a given mountain is nearly identical genetically to others. In this case, individuals in Population 1 will all be nearly identical to each other; individuals in Population 2 will also be identical to each other, but because populations are very isolated from one another, they may all be very different from all the individuals in Population 1; and so on. Under these circumstances, most of the variation in Species A is *among* populations; within-population variation is low or nonexistent.

Now imagine a contrary example. Species B lives in tallgrass prairie, where suitable habitats are relatively close to one another (or even continuous), and dispersal among populations is common. Here we might find a great deal of variation among individuals in each population (reflecting the benefits of variation discussed above), but because gene flow is high among sites, most populations are similar (that is, most polymorphic alleles are widely distributed). This illustrates a species with a high proportion of variation distributed *within* populations, while among-population variation is relatively small.

A variety of measures are used to quantify the distribution of genetic variation among individuals within populations, and among populations. These measures are the basis for describing how genetic variation is *partitioned* within species. Differences among populations are commonly quantified by the use of one of several statistics, including Wright's inbreeding coefficient (*FST*) and Nei's coefficient of gene variation (*GST*). These indices are functions of how heterozygosity is partitioned within and among populations, based on differences in allele frequencies. Where *pi* is the frequency of the *i*th allele, heterozygosity (*H*) = 1- Ó *pi*2 for {i = 1...n} for populations within a larger sample.

Then, the proportion of total variation that is distributed among populations (*GST*) is *HT- HSHT*, or equivalently, 1- (*HS/HT*), where *HS* and *HT* are the mean heterozygosity within populations and in the entire species respectively. Values for these statistics can often (although not often enough!) be found in the population genetic literature, so restorationists do not have to generate them. Values of *FST* and *GST* vary from 0 to 1.

GST in particular has a number of useful properties: it can be used for one or many loci, mutation rates do not alter the statistic significantly, the exact number of populations need not be specified, and the statistic is relatively responsive to changes in allele frequencies in time. Although they have important conceptual differences, in practice *FST* and *GST* are used in similar fashion as indices of genetic difference among populations (Crow 1986). Species genetic structure can be modelled in a variety of ways. The simplest is a "finite island" conceptual model, in which alleles can flow between any pair of populations. In this case, the key parameters are the migration rate and *effective population size* (Crow 1986). However, the assumption that any two populations are equally likely to exchange alleles is probably unrealistic. Instead, a "stepping stone" model can be used, in which alleles spread by "stepping" from one population to an adjacent one.

In practical terms, *G* and *F* statistics tell us whether the majority of genetic variation is distributed among or within populations. In species with low *GST* (approaching 0), the majority of variation is found *within populations*; individuals within populations are likely to be genetically different, but each population contains the same complement of alleles in similar frequencies Where *GST* is high (approaching 1), individuals within a population are relatively similar but *populations* are significantly different. Most species fall somewhere in between these extremes.

The distribution of genetic variation within and among species is strongly linked to life-history traits, particularly dispersal and *reproductive mode*. Species that disperse genes (in plants, this includes both pollen and seeds) widely and frequently will tend, other things being equal, to have lower *GST* (*i.e.*, populations will be more similar). Even a moderate rate of gene movement among populations (one individual every a few generations) can "link" the gene pools of two populations. Mutation and drift (chance selection of genotypes, especially likely in small populations) can also lead to changes in allele frequencies, although in general these forces are believed to act more slowly than dispersal and selection.

How is Genetic Variation Detected and Measured?

A variety of methods exist for the assessment of genetic variation. Traditionally, genetic variation was inferred by the shape (*morphology*) or growth responses of organisms. Such approaches often use *common garden experiments* or *reciprocal transplants* to distinguish genotypic and phenotypic variation. In this approach, individuals that appear phenotypically different (or that grow in different environments) are placed in a common environment. This eliminates the environmental component of variation; presumably, the remaining variation has a genetic basis. variation. While these methods can be useful, they measure higher-order effects of genetic variation (*i.e.*, expression in the whole organism), and hence address genetic variation only indirectly. Moreover, the confounding effects of phenotypic variation are often difficult to separate from underlying genetic differences among organisms. In practical terms, it is often not possible to know whether organisms that look different are actually different genetically, or whether their differing phenotypes reflect the influences of environment or chance in development.

Since the 1960's the most widely applied method of estimating genetic differences among individuals is *enzyme electrophoresis* (also referred to as *starch gel electrophoresis*, including both *allozyme* and *isozyme* analysis). Electrophoresis detects variation in amino acids that are early products of gene *translation* (the process by which a cell translates a copy of its genetic code in messenger-RNA into amino acids). Since there may be multiple biochemical pathways leading to the synthesis of a single amino acid or enzyme, electrophoresis is relatively conservative in its estimates of genetic variation among individuals. Enzyme electrophoresis remains the most widely used method of estimating genetic variation, and literally thousands of studies have been conducted on a wide range of organisms. Electrophoretic analysis permits individuals to be distinguished from one another based on enzyme variation. These differences among individuals can be compiled to generate statistics about the degree of variation within and among populations (§2.b), a key consideration in restoration ecology.

In the past decade, powerful methods have developed and emerged to detect variation in DNA itself, not simply its products. While a Introduction to restoration genetics. Falk, Knapp, and Guerrant, Society for Ecological Restoration, 17 December 2001. Bewildering variety of tools and techniques may be encountered in the literature, they all have in common the ability to reconstruct the sequence of nucleotides in the DNA molecule, arguably the most basic level of genetic variation. The conceptually simplest (although experimentally most laborious) technique is *gene sequencing* — that is, the exact reconstruction of the complete nucleotide sequence. Part of the difficulty in interpreting DNA sequences is that "genes" in a functional sense may be distributed over several sections of a chromosome; there are also large sections of all chromosomes that do not code for any known function, or that are redundant or *regulatory* (influence the transcription, translation, or function of other genes).

Thus, unlike enzyme analysis, much of the information in a DNA sequence may be of little relevance to understanding ecologically significant genetic variation. Since in theory every sexually-produced individual is likely to be unique, DNA sequences can be almost too variable to detect the kinds of patterns required for restoration work. However, new analytical techniques allow DNA sequence variation to be interpreted in meaningful ways. Complete gene sequences have been complied for relatively few organisms, and generally only a few individuals each, although these numbers are increasing all the time. It is important to note that isozymes and genetic markers may not reflect traits under strong natural selection. For example, Knapp and Rice ((1998) evaluated patterns of variation in a native grass using both quantitative traits and isozymes, and compared these to geographic distance and climate. They found overall that isozymes did not reflect an adaptation to climate, whereas the quantitative traits did. In a restoration context, this suggests that climate may in some cases serves as more useful guide for collecting and reintroduction zones that genetic markers *per se.*

How do Genetic Questions Enter into Restoration?

The genetic principles outlined above have a direct bearing on the practice of ecological restoration. In this section we summarize how restoration practitioners and researchers can (and should) take population genetics into account.

Accuracy and Functionality of Restored Populations

Sophisticated restoration practitioners recognize that the starting pool of genetic variation to be used is a critical element in design and implementation. However, opinions abound on how to select a suitable source pool for a restoration project. The current debate often confuses two related, but distinct issues: the *accuracy* ("authenticity") of a restoration project, and its *functionality* (Clewell, 2000). A reintroduction project is genetically *accurate* ("authentic") if it replicates the original gene pool of the population it replaces. If the original population has been destroyed, then perfect accuracy is strictly speaking impossible, because some alleles were probably unique to that population (*i.e.*, GST < 1). A restoration can be accurate and function poorly; it can also be inaccurate but functional. Since no reintroduction is perfectly accurate genetically, the question then becomes: "How close is close enough?" To this there can be no absolute ecological answer, since not all variation is of adaptive significance. Moreover, variation among populations is a continuous variable: two populations can differ by one allele or thousands. If the motivation is an ethical commitment to fidelity to the historic distribution of genotypes, the line can be drawn anywhere. The best practice is to estimate and report the degree of genetic accuracy based on available information, so that others studying a project can take authenticity into account.

How well a reintroduced population functions is a very different matter. Here the concern is not so much whether specific genotypes have been re-established, but rather how well a restored population will work in terms of persistence, resilience, and stability. Since many adaptive traits have a genetic basis, reintroduction material that performs (survives, grows, reproduces) well may come from a site with similar ecological conditions, but not necessarily close geographically. Moreover, if functionality is the main objective, then a range of genotypes can be introduced, allowing selection (differential survival and reproduction) to sort out the best for the site. If this approach is taken, there may be considerable attrition, which can be accounted for in initial sampling.

Geographic Location of Source Material for (Re) Introduction

Perhaps no single genetic issue is more intensely debated in restoration circles than the location of source populations: "Where should material for restoration come from?" Interest in population genetic variation often arises in the context of asking whence source material should come. In other words the restorationist is concerned not only with the degree of variability but its particular geographic distribution.

The most common approach is to specify a geographic (and often elevation) range, within which source material can be confidently collected. This is essentially a "space-for-genotype" substitution: we expect that populations near one another, and growing in similar conditions, will be more similar genetically due both to ecotypic variation and the effects of gene flow. Geographic distance is a reasonable first approximation of genetic distance, assuming that genetic diversity most likely obeys the "distance decay of similarity" in broad outline. In population genetic terms, this favours the "stepping stone" model of gene flow. On public lands in the western United States, commonly-applied guidelines suggest that material for outplanting be collected from within 1,000 ft. elevation bands and 100 miles lateral distance of the planting area (elevation bands vary from 500-3,000 ft for different agencies). As noted above climatic zones may also offer a better first approximation of genetic distance between populations than simple geographic distance.

Unfortunately, there are no simple distance rules that apply equally to all species, because species vary in gene flow among populations, population size, and the resulting distribution of diversity (GST). For some species (*e.g.*, self-fertilizing plants in small, isolated patches of habitat, or fishes in isolated stream reaches), each site may reflect a unique local adaptation, and the geographic range of suitable genotypes can be very small (a few km2). Other species (for example, species with wind dispersed pollen and seeds) with higher rates of gene flow, and those with larger effective populations, are probably less differentiated over the landscape and can be collected from much wider ranges. Most species have dispersal curves that are leptokurtic (*i.e.*, with long tails), meaning that a few seeds or offspring in each generation may travel far beyond the majority; while few in number, these long-range dispersers may play a critical role in helping species to adjust their ranges in periods of climate and vegetation change. For each species, the Introduction to restoration genetics.

Falk, Knapp, and Guerrant, Society for Ecological Restoration, 17 December 2001. Restorationist must ask: "How widely does this species disperse its genes under normal conditions, and what factors (distance, geographic barriers, habitat types) influence where genes can spread?" This uncertainty is reflected in the wide range of recommendations for suitable collecting zones, which range over more than three orders of magnitude from 100 meters to 100 miles.

The issue of geographic range for source material cannot be separated from the distinction made earlier between historical accuracy and functionality. If historical accuracy (authenticity) *sui generis* is the primary concern, then the range of potential collecting sites is governed most strongly by historical patterns of dispersal, and the collecting radius will be very small. If population function is the primary concern, on the other hand, then the net may be cast over a wider area (although still focusing on similar ecological communities). This is because populations that are widely separated in space may nonetheless occupy similar ecological settings, and by selection may have developed genotypes that are similar in key ecological traits. If the main goal is population function, then it is probably more important to derive source material from large, diverse populations than from small, genetically depauperate sites, even if the latter are closer geographically.

Of course, it is always possible to mix genotypes from different populations, and to let selection sort out the variation: this is, after all, exactly what nature does. Mixing of genotypes may be particularly suitable if existing "populations" are in fact isolated and reduced remnants of formerly widespread and interconnected groups. Re-combining fragmented populations genetically is appropriate as the geographic scale is ecologically realistic, and should not be done indiscriminately. Mixing too broadly (*i.e.*, combining genotypes from very diverse ecological settings) can result in too many individuals that are poorly adapted for the new given environment (high genetic load). Some geneticists advocate the use of regional mixtures — composite collections of genotypes, all of which are at least moderately well adapted to a given environment. before the population dies out.

There are valid arguments on both sides of the authentic *vs.* functionality debate. On one hand, there is little question that the gene pools of many remnant native populations have been seriously eroded, so that what occurs today is a small remnant of the original Introduction to restoration genetics. Falk, Knapp, and Guerrant, Society for Ecological Restoration, 17 December 2001, Diversity. Small gene pools are more prone to inbreeding, as well as random genetic change from drift. Populations that formerly exchanged genes regularly may have also become genetically isolated by habitat fragmentation. In such cases, a good argument can be made to bring together genetic material from several nearby populations, in effect replicating the natural (but now disrupted) processes of gene flow. In addition, some restoration sites may be so heavily disturbed (*i.e.* mine tailing reclamation areas) that the most "local" population may no longer be the one best adapted to the growing environment. In such cases, a wide diversity of genotypes can increase the chances that at least some plants will have the genetic composition to survive..

It is possible, however, that genotypes from outside the apparent current range may perform too well — that is, swamp the local genotype. If remnant native populations of the species being introduced already exist at a site, additional genetic considerations reinforce the importance of the scale of collection. Introduced populations may hybridize with the existing native population, introducing new genes (genetic pollution) and potentially negatively affecting genetic integrity. If a few poorly adapted individuals make it into the existing native population, it might be argued that natural selection will eventually remove the deleterious genes. However, the issue is partly one of numbers; with commercial seed production of many native species, restorationists now have the tools to dump large volumes of seed into ecosystems. If the number of poorly adapted, non-local propagules is large in relation to the number of local native types, the chance of matings between the well adapted and the poorly adapted plants will increase (especially in cross-pollinating species), thereby potentially swamping the native population and diluting the local genes. Progeny from such matings may additionally experience outbreeding depression (*i.e.* poor survival and growth in relation to the parents). Carefully chosen introductions, with genotypes similar to the existing native population, can avoid these negative impacts.

Take-home messages: Since no single rule applies to all situations, the most important recommendation is to use explicit and reasoned criteria in selecting source populations. To summarize the issues presented here: Introduction to restoration genetics. Falk, Knapp, and Guerrant, Society for Ecological Restoration, 17 December 2001.

1. Species vary in their dispersal rates and distances, and hence in the degree of genetic differentiation among populations. These differences are often closely correlated with life-history attributes particular to each species that affect rates and patterns of gene flow. Therefore, guidelines are inherently species-specific.
2. The restorationist must decide if historical accuracy or functionality is the primary objective. True historical accuracy can be difficult to determine if existing populations are remnants of a formerly widespread range. Moreover, most natural populations experience a degree of genetic change over time and space. However, even where the emphasis is on functionality, historical reference conditions are essential to "anchor" restoration work within a natural range of variability.

3. Geographic distance (*i.e.*, the "space-for-genotype" substitution) is a reasonable, but crude, substitute for patterns of gene flow among populations. However, in a disrupted landscape the most local remnant populations may be genetically reduced, and may not have genotypes that correspond to the conditions under restoration. Moreover, if populations are strongly selected to local habitat conditions, then habitat similarity may outweigh distance as a selection criterion. Local and regional climatic and soil zones can be useful criteria for obtaining material for reintroduction, since these will often represent plants adapted to the conditions at the restoration site.
4. As a source of material for reintroduction, large, genetically diverse source populations are generally preferable to small, limited populations even when the latter are closer to the restoration site. It may be preferable to combine material from several suitable sites, to capture a wider array of genotypes that can succeed in the new location. Such "regional mixtures" should draw on populations within a similar ecological zone (climate, soil) to the restoration site, to avoid too large a proportion of individuals that may not be well adapted to the new conditions.
5. Small local populations can be "swamped" by non-native genotypes if a species already exists at a restoration site. If existing populations are to be Introduction to restoration genetics. Falk, Knapp, and Guerrant, Society for Ecological Restoration, 17 December 2001, augmented, the number of individuals introduced from other locations should not be so large as to overwhelm the local gene pool, particularly if there is evidence of local adaptation.

Sampling the Diversity of Source Populations

The initial gene pool for a restoration project is inevitably limited by the diversity of the original sample. While other genes may enter the project area over time (for instance, by migration of individuals, dispersal of gametes, or additional reintroduction measures), the starting pool of genetic diversity will govern the performance of a reintroduced population for a long time. Hence, the diversity of the original source collection is a critical consideration in restoration.

A variety of guidelines have been developed for sampling wild populations of plants and animals for breeding and reintroduction (Centre for Plant Conservation 1991; Guerrant 1992; BGCI 1994; Guerrant 1996; IUCN/SSC 1998). These guidelines address four fundamental sampling issues:

1. How many populations will be sampled to create the source pool?
2. How many individuals will be sampled from each population?

3. Will samples be collected all at once, or over a period of years?
4. What is the probability of a collected sample surviving to establishment?

Number of populations: In most species, the cumulative amount of genetic variation captured increases as successive populations are added to the sample. However, since populations have some measure of similarity (0 < GST < 1), each additional population added to a sample collects some alleles that are new to the sample, and some that are already present from previous samples. As the number of populations sampled increases, the *marginal diversity rate* decreases (that is, fewer and fewer novel alleles are captured), and the *cumulative diversity function* approaches an asymptote. For a pool of populations sampled at random, there comes a point at which further sampling across populations provides little or no additional genetic benefit.

The number of populations at which this occurs is related strongly to the measure of differentiation among populations, GST, which we have already encountered. When GST is high, populations are markedly different from one another, and more should be sampled Introduction to restoration genetics. Falk, Knapp, and Guerrant, Society for Ecological Restoration, 17 December 2001, to capture the maximum total diversity. When GST is low, most populations are similar, and sampling from only a few will capture most of the diversity that exists(provided that the statistic was generated using populations across the full range of the species).. Of course, it is always worth remembering that a significant fraction of most genomes is monomorphic, and is therefore captured in the first population, if not the first individual.

There can be no hard and fast rules for the number of populations to be sampled across all taxa. In a recent survey of 13 plant species, the sample for some species was saturated for common isozyme alleles after just 3-5 populations (20% of the total), while in other species the cumulative diversity curve was still rising after all populations (as many as 15-30) were included in the sample (recall, however, that isozyme loci may be less variable among populations than other loci under strong selection). (Falk and Gibbs *In prep.*). Rare or uncommon alleles accumulate in the sample at a slower rate than do common alleles, so if it is suspected that important adaptations occur at low frequency (for instance, because of recent environmental change), then it is advisable to sample a larger number of sites. Further research is ongoing in this area; in the meantime, GST or some comparable measure of population differentiation is a suitable guide for the sampling strategy.

One exception to this general null model is for populations that are strongly differentiated along habitat lines (*e.g. ecotypes*). If the reintroduction area is unusual habitat for a species, then it is worth sampling any populations that occur in similar conditions, to increase the likelihood of capturing alleles that offer adaptive benefit.

Number of individuals to sample within populations. The underlying theoretical basis for sampling multiple individuals within a population is that populations are rarely truly *panmictic* (that is, with completely randomized breeding). In plants, a large proportion of mating occurs between neighboring individuals, even when pollination occurs via an animal vector. In animal populations, a wide range of behavioral adaptations exists that commonly concentrates breeding success in a few individuals at any given time. The result is that populations are not genetically homogeneous; to capture their genetic diversity adequately, multiple individuals need to be sampled.

Again, a number of general guidelines have been released on this topic, and again, there can be no hard and fast "magic numbers" that apply across taxa. Brown and Briggs 1991). Plant studies evaluated by the Centre for Plant Conservation and its colleague institutions suggest that anywhere from 10-50 individuals should be sampled per population, with the understanding that there is bound to be considerable redundancy among these samples. Sampling of propagules (seeds, vegetative shoots) has less impact on the source population, although there is often lower success associated with these life stages compared to outplanting established plants.

Number of Years During which Samples will be Collected

For many species, an adequate sample can be collected in a single year. However, for many others, collecting may need to be distributed over a number of years. Reasons for multiple-year collecting include:

- Variable reproductive output (masting), where large numbers of individuals are reproductive or large numbers of offspring are produced only in certain years;
- Small populations or species with low reproductive rates, where it is important to avoid depleting the pool of new individuals;
- Annual organisms, for which gene pools may be significantly different from year to year, either by selection or by chance; and
- Limited capacity to handle (store, propagate, release) collected material.

Probability of a Collected Sample Surviving to Establishment.

In the end, what counts in a reintroduction is the number of individuals in the new population as well as their diversity (Menges 1991). However, less than 100% of the samples (seeds, cutting, eggs, adults) collected in the field will survive to establishment. Attrition occurs at every step along the way: during the collecting process, transportation, storage, propagation/curation, and outplanting/release. High initial mortality rates are often observed in reintroduced populations, often continuing for several years. A simple calculation can help to account for attrition of collected samples. Let *Ps* represent the survival probability of a collected sample s, and N the number of individuals desired in the final restored population. Then we account for attrition during the reintroduction process by collecting PsN samples in the field.

Genetic Diversity within Reintroduced Populations

As discussed above, breeding among closely related individuals can lead to expression of unfavourable traits that compromise an organism's ability to survive and reproduce. Individuals resulting from inbreeding in populations where this did not occur naturally from highly inbred lines can have stunted growth, altered behaviour, deformed morphology, poor physiological function; such individuals are also more likely to be reproductively sterile.

Hence, the genetic origin of individuals used in a restoration project can influence whether a project succeeds or fails. If source collections are made from diverse, naturally occurring populations, the probability of breeding among close relatives is reduced. A more diverse population may include individuals that can tolerate a wider range of conditions. If, on the other hand, all individuals in a restored population are genetically similar, and the total population size is not large (fewer than 100 individuals), the probability is great that their offspring will be homozygous at some loci, and thus have reduced fitness. A genetically narrow population may be able to survive only in a narrow range of conditions. While genetic diversity is not always associated with *ecological amplitude*, in general the correlation is positive (Huenneke 1991).

In addition to affecting how well a restored population succeeds in current environmental conditions, genetic diversity may affect performance over time. Important ecological traits, such as tolerance of disturbance or climatic extremes often have a strong genetic basis. Such traits need to be considered not only in terms of conditions at the time of reintroduction, but in light of the natural range of variability over many years or generations. If the goal of restoration is to establish self-sustaining populations (Pavlik 1996), then the range of conditions to which the population and its offspring will be exposed is a vital consideration. Genetically uniform populations may do well one year when conditions are favourable for their genotype, and then fail the next, even though the range of environmental variation is within normal limits. Such failures are even more Introduction to restoration genetics.

Falk, Knapp, and Guerrant, Society for Ecological Restoration, 17 December 2001, likely in the case of less common extreme events (Gaines and Denny 1993). Recalling from § 1.b and 1.d that environmental conditions vary within an envelope of natural variation, it is safe to assume that most reintroduced populations will be exposed to a wide range of conditions over time. Hence, the genetic diversity of reintroduced populations is a non-trivial consideration for their long-term persistence (Montalvo et al. 1997). It is not uncommon in restoration work for large numbers of individuals to be released or outplanted, a large proportion of which fail to survive beyond 2-3 years. This means that restoration releases are subject both to the *founder effect* (the result of the initially limited gene pool) and then further *genetic bottlenecks* as effective population size diminishes due to mortality. If the initial outplanting or release is itself genetically uniform (*e.g.*, hundreds or thousands of clonally-produced plants), the resulting gene pool in the field can be quite narrow. Such apparently large populations are, in genetic terms, very small.

Of course, genetically narrow populations occur in nature, often resulting from similar forces (founder events and bottlenecks). For example, aspen (*Populus tremuloides*) and many species in the Iridaceae (*Iris*, *Lilium*) are often found in large clumps of many *ramets* of very few *genets*, or even just one. Natural populations of species where sexual (as opposed to vegetative or parthenogenic) reproduction dominates are less likely to be genetically homogeneous (Meffe and Carroll 1994). The effects of genetic homogeneity in a reintroduced population may not be immediately evident, but over a period of years the population may have lower rates of growth, survival and reproduction, and may persist less successfully through periods of natural environmental variability.

What you can do: Restorationists should be aware of the level of genetic diversity that they are working with. Above all, restoration planners and managers should understand how the plants and animals they use were generated. Ask your nursery or whoever collected and propagated your seed source (or your breeder, in the case of animals) what methods were used, and what steps were taken to assure the presence of a suitably wide range of genotypes. While clonally-produced flats of thousands of identical plants may appear to offer short-term advantages (as they do for agricultural crops) of Introduction to restoration genetics. Falk, Knapp, and Guerrant, Society for Ecological Restoration, 17 December 2001, predictable response to current growing conditions, in the long run (again as with crops) such populations may be less likely to persist and succeed in the face of disease, competition and climate variability. Of course, knowing the methods by which individuals or material was produced provides genetic information only by inference; it is unfortunately not common to have good genetic data for reintroduction efforts.

Genetic Diversity Among Reintroduced Populations

The same principles apply to diversity among population diversity. Most species have a degree of differentiation among populations (*i.e.*, GST > 0). Population geneticists debate whether these differences have adaptive value in any given case – that is, whether they are the result of selection – or whether they represent the effects of sampling error (founder effects, drift, mutation). Nonetheless, a reasonable precautionary position is to mimic the degree of genetic difference found in natural populations of a species when (re)introducing populations to the landscape.

Most species, however, have little or no published data on their among-population genetic diversity. The few surveys that have been done are invaluable sources of information, and restorationists should read and cite them whenever possible. In some cases studies of *congeners* can be used, although with the caveat that species that are closely related phylogenetically are not always similar ecologically.

Resources

A vast literature exists on genetic diversity, its distribution in space and time, and its adaptive and ecological significance. A number of excellent texts on population genetics provide greater depth on a number of topics addressed briefly in this paper. Several of the growing numbers of papers on the topic of restoration genetics are listed in the literature cited section of this paper. Conservation genetics is still is relatively small field, and restoration genetics is in its infancy. Restoration practitioners and scientists can contribute to the accumulation of knowledge in this area by treating your restoration projects as experiments, and publishing your results. The Society for Ecological Restoration maintains an on-line *Directory of Restoration Expertise* at www.ser.org. The IUCN also has published a list of Introduction to restoration genetics. Falk, Knapp, and Guerrant, Society for Ecological Restoration, 17 December 2001, reintroduction practitioners (Soorae and Seddon 1998).

How Genes Work

People have known for many years that living things inherit traits from their parents. That common-sense observation led to agriculture, the purposeful breeding and cultivation of animals and plants for desirable characteristics. Firming up the details took quite some time, though. Researchers did not understand exactly how traits were passed to the next generation until the middle of the 20th century. Now it is clear that genes are what carry our traits through generations and that genes are made of deoxyribonucleic acid (DNA). But genes themselves don't do the actual work. Rather they serve as instruction books for making functional molecules such as ribonucleic acid (RNA) and proteins, which perform the chemical reactions in our bodies. Proteins do many other things, too. They provide the body's main building materials, forming the cell's architecture and structural components. But one thing proteins can't do is make copies of themselves. When a cell needs more proteins, it uses the manufacturing instructions coded in DNA.

The DNA code of a gene—the sequence of its individual DNA building blocks, labeled A (adenine), T (thymine), C (cytosine), and G (guanine) and collectively called nucleotides— spells out the exact order of a protein's building blocks, amino acids. Occasionally, there is a kind of typographical error in a gene's DNA sequence. This mistake— which can be a change, gap, or duplication—is called a mutation. A mutation can cause a gene to encode a protein that works incorrectly or that doesn't work at all. Sometimes, the error means that no protein is made. But not all DNA changes are harmful. Some mutations have no effect, and others produce new versions of proteins that may give a survival advantage to the organisms that have them. Over time, mutations supply the raw material from which new life forms evolve.

What are Proteins and what do they Do?

Proteins are large, complex molecules that play many critical roles in the body. They do most of the work in cells and are required for the structure, function, and regulation of the body's tissues and organs. Proteins are made up of hundreds or thousands of smaller units called amino acids, which are attached to one another in long chains. There are 20 different types of amino acids that can be combined to make a protein. The sequence of amino acids determines each protein's unique 3-dimensional structure and its specific function.

How do Genes Direct the Production of Proteins?

Most genes contain the information needed to make functional molecules called proteins. (A few genes produce other molecules that help the cell assemble proteins.) The journey from gene to protein is complex and tightly controlled within each cell. It consists of two major steps: transcription and translation. Together, transcription and translation are known as gene expression.

During the process of transcription, the information stored in a gene's DNA is transferred to a similar molecule called RNA (ribonucleic acid) in the cell nucleus. Both RNA and DNA are made up of a chain of nucleotide bases, but they have slightly different chemical properties. The type of RNA that contains the information for making a protein is called messenger RNA (mRNA) because it carries the information, or message, from the DNA out of the nucleus into the cytoplasm.

Translation, the second step in getting from a gene to a protein, takes place in the cytoplasm. The mRNA interacts with a specialized complex called a ribosome, which "reads" the sequence of mRNA bases.

Each sequence of three bases, called a codon, usually codes for one particular amino acid. (Amino acids are the building blocks of proteins.) A type of RNA called transfer RNA (tRNA) assembles the protein, one amino acid at a time. Protein assembly continues until the ribosome encounters a "stop" codon (a sequence of three bases that does not code for an amino acid).

Can Genes be Turned on and Off in Cells?

Each cell expresses, or turns on, only a fraction of its genes. The rest of the genes are repressed, or turned off. The process of turning genes on and off is known as gene regulation. Gene regulation is an important part of normal development. Genes are turned on and off in different patterns during development to make a brain cell look and act different from a liver cell or a muscle cell, for example. Gene regulation also allows cells to react quickly to changes in their environments. Although we know that the regulation of genes is critical for life, this complex process is not yet fully understood.

Gene regulation can occur at any point during gene expression, but most commonly occurs at the level of transcription (when the information in a gene's DNA is transferred to mRNA). Signals from the environment or from other cells activate proteins called transcription factors. These proteins bind to regulatory regions of a gene and increase or decrease the level of transcription. By controlling the level of transcription, this process can determine the amount of protein product that is made by a gene at any given time.

How do Cells Divide?

There are two types of cell division: mitosis and meiosis. Most of the time when people refer to "cell division," they mean mitosis, the process of making new body cells. Meiosis is the type of cell division that creates egg and sperm cells. Mitosis is a fundamental process for life. During mitosis, a cell duplicates all of its contents, including its chromosomes, and splits to form two identical daughter cells. Because this process is so critical, the steps of mitosis are carefully controlled by a number of genes. When mitosis is not regulated correctly, health problems such as cancer can result. The other type of cell division, meiosis, ensures that humans have the same number of chromosomes in each generation. It is a two-step process that reduces the chromosome number by half—from 46 to 23—to form sperm and egg cells. When the sperm and egg cells unite at conception, each contributes 23 chromosomes so the resulting embryo will have the usual 46. Meiosis also allows genetic variation through a process of DNA shuffling while the cells are dividing.

How do Genes Control the growth and Division of Cells?

A variety of genes are involved in the control of cell growth and division. The cell cycle is the cell's way of replicating itself in an organized, step-by-step fashion. Tight regulation of this process ensures that a dividing cell's DNA is copied properly, any errors in the DNA are repaired, and each daughter cell receives a full set of chromosomes. The cycle has checkpoints (also called restriction points), which allow certain genes to check for mistakes and halt the cycle for repairs if something goes wrong. If a cell has an error in its DNA that cannot be repaired, it may undergo programmed cell death (apoptosis) (illustration). Apoptosis is a common process throughout life that helps the body get rid of cells it doesn't need. Cells that undergo apoptosis break apart and are recycled by a type of white blood cell called a macrophage (illustration). Apoptosis protects the body by removing genetically damaged cells that could lead to cancer, and it plays an important role in the development of the embryo and the maintenance of adult tissues. Cancer results from a disruption of the normal regulation of the cell cycle. When the cycle proceeds without control, cells can divide without order and accumulate genetic defects that can lead to a cancerous tumor (illustration).

How do Geneticists Indicate the Location of a Gene?

Geneticists use maps to describe the location of a particular gene on a chromosome. One type of map uses the cytogenetic location to describe a gene's position. The cytogenetic location is based on a distinctive pattern of bands created when chromosomes are stained with certain chemicals. Another type of map uses the molecular location, a precise description of a gene's position on a chromosome. The molecular location is based on the sequence of DNA building blocks (base pairs) that make up the chromosome.

Cytogenetic Location

Geneticists use a standardized way of describing a gene's cytogenetic location. In most cases, the location describes the position of a particular band on a stained chromosome:

17q12

It can also be written as a range of bands, if less is known about the exact location:

17q12-q21

The combination of numbers and letters provide a gene's "address" on a chromosome. This address is made up of several parts:

- The chromosome on which the gene can be found. The first number or letter used to describe a gene's location represents the chromosome. Chromosomes 1 through 22 (the autosomes) are designated by their chromosome number. The sex chromosomes are designated by X or Y.
- The arm of the chromosome. Each chromosome is divided into two sections (arms) based on the location of a narrowing (constriction) called the centromere. By convention, the shorter arm is called p, and the longer arm is called q. The chromosome arm is the second part of the gene's address. For example, 5q is the long arm of chromosome 5, and Xp is the short arm of the X chromosome.
- The position of the gene on the p or q arm. The position of a gene is based on a distinctive pattern of light and dark bands that appear when

the chromosome is stained in a certain way. The position is usually designated by two digits (representing a region and a band), which are sometimes followed by a decimal point and one or more additional digits (representing sub-bands within a light or dark area). The number indicating the gene position increases with distance from the centromere. For example: 14q21 represents position 21 on the long arm of chromosome 14. 14q21 is closer to the centromere than 14q22.

Sometimes, the abbreviations "cen" or "ter" are also used to describe a gene's cytogenetic location. "Cen" indicates that the gene is very close to the centromere. For example, 16pcen refers to the short arm of chromosome 16 near the centromere. "Ter" stands for terminus, which indicates that the gene is very close to the end of the p or q arm. For example, 14qter refers to the tip of the long arm of chromosome 14. ("Tel" is also sometimes used to describe a gene's location. "Tel" stands for telomeres, which are at the ends of each chromosome. The abbreviations "tel" and "ter" refer to the same location.)

Molecular Location

The Human Genome Project, an international research effort completed in 2003, determined the sequence of base pairs for each human chromosome. This sequence information allows researchers to provide a more specific address than the cytogenetic location for many genes. A gene's molecular address pinpoints the location of that gene in terms of base pairs. It describes the gene's precise position on a chromosome and indicates the size of the gene. Knowing the molecular location also allows researchers to determine exactly how far a gene is from other genes on the same chromosome. Different groups of researchers often present slightly different values for a gene's molecular location. Researchers interpret the sequence of the human genome using a variety of methods, which can result in small differences in a gene's molecular address. Genetics Home Reference presents data from NCBI for the molecular location of genes.

What are Gene Families?

A gene family is a group of genes that share important characteristics. In many cases, genes in a family share a similar sequence of DNA building blocks (nucleotides). These genes provide instructions for making products (such as proteins) that have a similar structure or function. In other cases, dissimilar genes are grouped together in a family because proteins produced from these genes work together as a unit or participate in the same process.

Classifying individual genes into families helps researchers describe how genes are related to each other. Researchers can use gene families to predict the function of newly identified genes based on their similarity to known genes. Similarities among genes in a family can also be used to predict where and when a specific gene is active (expressed). Additionally, gene families may provide clues for identifying genes that are involved in particular diseases. Sometimes not enough is known about a gene to assign it to an established family. In other cases, genes may fit into more than one family. No formal guidelines define the criteria for grouping genes together. Classification systems for genes continue to evolve as scientists learn more about the structure and function of genes and the relationships between them. Genetics Home Reference provides information about gene families including a brief description of each gene family and a list of the genes included in the family.

The HUGO Gene Nomenclature Committee (HGNC) has classified many human genes into families. Each grouping is given a name and symbol, and contains a table of the genes in that family. The textbook Human Molecular Genetics (second edition, 1999) provides background information on human gene families. The Gene Ontology database lists the protein products of genes by their location within the cell (cellular component), biological process, and molecular function. The Reactome database classifies the protein products of genes based on their participation in specific biological pathways. For example, this resource provides tables of genes involved in controlled cell death (apoptosis), cell division, and DNA repair.

Understanding Gene Testing

Genes- the chemical messages of heredity- constitute a blueprint of our possibilities and limitations. The legacy of generations of ancestors, our genes carry the key to our similarities and our uniqueness. When genes are working properly, our bodies develop and function smoothly. But should a single gene- even a tiny segment of a single gene- go awry, the effect can be dramatic: deformities and disease, even death. In the past 20 years, amazing new techniques have allowed scientists to learn a great deal about how genes work and how genes are linked to disease. Increasingly, researchers are able to identify mutations, changes within genes that can lead to specific disorders. Tests for gene mutations make it possible not only to detect diseases already in progress but also, in certain situations, to foresee diseases yet to come.

This new ability raises both high hopes and grave concerns. On the one hand, predictive gene testing holds out the possibility of saving thousands of lives through prevention or early detection. On the other, the implications of test results are enormous, not only for the individual but also for relatives who share this genetic legacy, and for society as a whole. This booklet presents key concepts and issues relevant to gene testing and answers questions that are frequently asked about the techniques, potential risks, and possible benefits of attempts to link genetic markers with disease. Words that appear underlined in the pages that follow are defined in the Glossary. Clicking on these words will take you to the glossary. From there press the browser's back button to return.

What are Genes?

Genes are working subunits of DNA. DNA is a vast chemical information database that carries the complete set of instructions for making all the proteins a cell will ever need. Each gene contains a particular set of instructions, usually coding for a particular protein. DNA exists as two long, paired strands spiraled into the famous double helix. Each strand is made up of millions of chemical building blocks called bases. While there are only four different chemical bases in DNA (adenine, thymine, cytosine, and guanine), the order in which the bases occur determines the information available, much as specific letters of the alphabet combine to form words and sentences.

DNA resides in the core, or nucleus, of each of the body's trillions of cells. Every human cell (with the exception of mature red blood cells, which have no nucleus) contains the same DNA. Each cell has 46 molecules of double-stranded DNA. Each molecule is made up of 50 to 250 million bases housed in a chromosome. The DNA in each chromosome constitutes many genes (as well as vast stretches of noncoding DNA, the function of which is unknown). A gene is any given segment along the DNA that encodes instructions that allow a cell to produce a specific product- typically, a protein such as an enzyme- that initiates one specific action. There are between 50,000 and 100,000 genes, and every gene is made up of thousands, even hundreds of thousands, of chemical bases.

How do Genes Work?

Although each cell contains a full complement of DNA, cells use genes selectively. Some genes enable cells to make proteins needed for basic functions; dubbed housekeeping genes, they are active in many types of cells. Other genes, however, are inactive most of the time. Some genes play a role in early development of the embryo and are then shut down forever. Many genes encode proteins that are unique to a particular kind of cell and that give the cell its character- making a brain cell, say, different from a bone cell. A normal cell activates just the genes it needs at the moment and actively suppresses the rest.

RNA and DNA Revealed: New Roles, New Rules

For many years, when scientists thought about heredity, DNA was the first thing to come to mind. It's true that DNA is the basic ingredient of our genes and, as such, it often steals the limelight from RNA, the other form of genetic material inside our cells. But, while they are both types of genetic material, RNA and DNA are rather different. The chemical units of RNA are like those of DNA, except that RNA has the nucleotide uracil (U) instead of thymine (T). Unlike double-stranded DNA, RNA usually comes as only a single strand. And the nucleotides in RNA contain ribose sugar molecules in place of deoxyribose. RNA is quite flexible—unlike DNA, which is a rigid, spiral-staircase molecule that is very stable. RNA can twist itself into a variety of complicated, three-dimensional shapes. RNA is also unstable in that cells constantly break it down and must continually make it fresh, while DNA is not broken down often. RNA's instability lets cells change their patterns of protein synthesis very quickly in response to what's going on around them.

Many textbooks still portray RNA as a passive molecule, simply a "middle step" in the cell's gene-reading activities. But that view is no longer accurate. Each year, researchers unlock new secrets about RNA. These discoveries reveal that it is truly a remarkable molecule and a multi-talented actor in heredity. Today, many scientists believe that RNA evolved on the Earth long before DNA did. Researchers hypothesize—obviously, no one was around to write this down—that RNA was a major participant in the chemical reactions that ultimately spawned the first signs of life on the planet.

RNA World

At least two basic requirements exist for making a cell: the ability to hook molecules together and break them apart, and the ability to replicate, or copy itself, from existing information. RNA probably helped to form the first cell. The first organic molecules, meaning molecules containing carbon, most likely arose out of random collisions of gases in the Earth's primitive atmosphere, energy from the Sun, and heat from naturally occurring radioactivity. Some scientists think that in this primitive world, RNA was a critical molecule because of its ability to lead a double life: to store information and to conduct chemical reactions. In other words, in this world, RNA served the functions of both DNA and proteins.

What does any of this have to do with human health? Plenty, it turns out. Today's researchers are harnessing some of RNA's flexibility and power. For example, through a strategy he calls directed evolution, molecular engineer Ronald R. Breaker of Yale University is developing ways to create entirely new forms of RNA and DNA that both work as enzymes. Recently, Breaker and others have also uncovered a hidden world of RNAs that play a major role in controlling gene activity, a job once thought to be performed exclusively by proteins. These RNAs, which the scientists named riboswitches, are found in a wide variety of bacteria and other organisms. This discovery has led Breaker to speculate that new kinds of antibiotic medicines could be developed to target bacterial riboswitches.

Molecular Editor

Scientists are learning of another way to customize proteins: by RNA editing. Although DNA sequences spell out instructions for producing RNA and proteins, these instructions aren't always followed precisely. Editing a gene's mRNA, even by a single chemical letter, can radically change the resulting protein's function. Nature likely evolved the RNA editing function as a way to get more proteins out of the same number of genes. For example, researchers have found that the mRNAs for certain proteins important for the proper functioning of the nervous system are particularly prone to editing. It may be that RNA editing gives certain brain cells the capacity to react quickly to a changing environment.

Which molecules serve as the editor and how does this happen? Brenda Bass of the University of Utah School of Medicine in Salt Lake City studies one particular class of editors called adenosine deaminases. These enzymes "retype" RNA letters at various places within an mRNA transcript. They do their job by searching for characteristic RNA shapes. Telltale twists and bends in folded RNA molecules signal these enzymes to change the RNA sequence, which in turn changes the protein that gets made. Bass' experiments show that RNA editing occurs in a variety of organisms, including people. Another interesting aspect of editing is that certain disease-causing microorganisms, such as some forms of parasites, use RNA editing to gain a survival edge when living in a human host. Understanding the details of this process is an important area of medical research.

Small But Powerful

Recently, molecules called microRNAs have been found in organisms as diverse as plants, worms, and people. The molecules are truly "micro," consisting of only a few dozen nucleotides, compared to typical human mRNAs that are a few thousand nucleotides long. What's particularly interesting about microRNAs is that many of them arise from DNA that used to be considered merely filler material. How do these small but important RNA molecules do their work? They start out much bigger but get trimmed by cellular enzymes, including one aptly named Dicer. Like tiny pieces of Velcro®, microRNAs stick to certain mRNA molecules and stop them from passing on their protein-making instructions.

First discovered in a roundworm model system some microRNAs help determine the organism's body plan. In their absence, very bad things can happen. For example, worms engineered to lack a microRNA called let-7 develop so abnormally that they often rupture and practically break in half as the worm grows. Perhaps it is not surprising that since microRNAs help specify the timing of an organism's developmental plan, the appearance of the microRNAs themselves is carefully timed inside a developing organism. Biologists, including Amy Pasquinelli of the University of California, San Diego, are currently figuring out how microRNAs are made and cut to size, as well as how they are produced at the proper time during development. MicroRNA molecules also have been linked to cancer. For example, Gregory

Hannon of the Cold Spring Harbor Laboratory on Long Island, New York, found that certain microRNAs are associated with the severity of the blood cancer B-cell lymphoma in mice. Since the discovery of microRNAs in the first years of the 21st century, scientists have identified hundreds of them that likely exist as part of a large family with similar nucleotide sequences. New roles for these molecules are still being found.

RNA Interference (RNAi)

Healthy Interference

RNA controls genes in a way that was only discovered recently: a process called RNA interference, or RNAi. Although scientists identified RNAi less than 10 years ago, they now know that organisms have been using this trick for millions of years. Researchers believe that RNAi arose as a way to reduce the production of a gene's encoded protein for purposes of fine-tuning growth or self-defence. When viruses infect cells, for example, they command their host to produce specialized RNAs that allow the virus to survive and make copies of itself. Researchers believe that RNAi eliminates unwanted viral RNA, and some speculate that it may even play a role in human immunity. Oddly enough, scientists discovered RNAi from a failed experiment! Researchers investigating genes involved in plant growth noticed something strange: When they tried to turn petunia flowers purple by adding an extra "purple" gene, the flowers bloomed white instead.

This result fascinated researchers, who could not understand how adding genetic material could somehow get rid of an inherited trait. The mystery remained unsolved until, a few years later, two geneticists studying development saw a similar thing happening in lab animals. The researchers, Andrew Z. Fire, then of the Carnegie Institution of Washington in Baltimore and now at Stanford University, and Craig Mello of the University of Massachusetts Medical School in Worcester, were trying to block the expression of genes that affect cell growth and tissue formation in roundworms, using a molecular tool called antisense RNA.

To their surprise, Mello and Fire found that their antisense RNA tool wasn't doing much at all. Rather, they determined, a doublestranded contaminant produced during the synthesis of the single-stranded antisense RNA interfered with gene expression. Mello and Fire named the process RNAi, and in 2006 were awarded the Nobel Prize in physiology or medicine for their discovery. Further experiments revealed that the doublestranded RNA gets chopped up inside the cell into much smaller pieces that stick to mRNA and block its action, much like the microRNA pieces of Velcro discussed above.

Today, scientists are taking a cue from nature and using RNAi to explore biology. They have learned, for example, that the process is not limited to worms and plants, but operates in humans too. Medical researchers are currently testing new types of RNAi-based drugs for treating conditions such as macular degeneration, the leading cause of blindness, and various infections, including those caused by HIV and herpes virus.

Dynamic DNA

A good part of who we are is "written in our genes," inherited from Mom and Dad. Many traits, like red or brown hair, body shape, and even some personality quirks, are passed on from parent to offspring. But genes are not the whole story. Where we live, how much we exercise, what we eat: These and many other environmental factors can all affect how our genes get expressed. You know that changes in DNA and RNA can produce changes in proteins. But additional control happens at the level of DNA, even though these changes do not alter DNA directly. Inherited factors that do not change the DNA sequence of nucleotides are called epigenetic changes, and they too help make each of us unique.

Epigenetic means, literally, "upon" or "over" genetics. It describes a type of chemical reaction that can alter the physical properties of DNA without changing its sequence. These changes make genes either more or less likely to be expressed. Currently, scientists are following an intriguing course of discovery to identify epigenetic factors that, along with diet and other environmental influences, affect who we are and what type of illnesses we might get.

Secret Code

DNA is spooled up compactly inside cells in an arrangement called chromatin. This packaging is critical for DNA to do its work. Chromatin consists of long strings of DNA spooled around a compact assembly of proteins called histones. One of the key functions of chromatin is to control access to genes, since not all genes are turned on at the same time. Improper expression of growth-promoting genes, for example, can lead to cancer, birth defects, or other health concerns.

Many years after the structure of DNA was determined, researchers used a powerful device known as an electron microscope to take pictures of chromatin fibres. Upon viewing chromatin up close, the researchers described it as "beads on a string," an image still used today. The beads were the histone balls, and the string was DNA wrapped around the histones and connecting one bead to the next. Decades of study eventually revealed that histones have special chemical tags that act like switches to control access to the DNA. Flipping these switches, called epigenetic markings, unwinds the spooled DNA so the genes can be transcribed. The observation that a cell's gene-reading machinery tracks epigenetic markings led C. David Allis, who was then at the University of Virginia Health Sciences Centre in Charlottesville and now works at the Rockefeller University in New York City, to coin a new phrase, the "histone code." He and others believe that the histone code plays a major role in determining which proteins get made in a cell. Flaws in the histone code have been associated with several types of cancer, and researchers are actively pursuing the development of medicines to correct such errors.

Genetics and You: The Genetics of Anticipation

Occasionally, unusual factors influence whether or not a child will be born with a genetic disease. An example is the molecular error that causes Fragile X syndrome, a rare condition associated with mental retardation. The mutation leading to a fragile X chromosome is not a typical DNA typing mistake, in which nucleotides are switched around or dropped, or one of them is switched for another nucleotide. Instead, it is a kind of stutter by the DNA polymerase enzyme that copies DNA. This stutter creates a string of repeats of a DNA sequence that is composed of just three nucleotides, CGG.

The number of triplet repeats seems to increase as the chromosome is passed down through several generations. Thus, the grandsons of a man with a fragile X chromosome, who is not himself affected, have a 40 percent risk of retardation if they inherit the repeat-containing chromosome. The risk for great-grandsons is even higher: 50 percent. Intrigued by the evidence that triplet repeats can cause genetic disease, scientists have searched for other examples of disorders associated with the DNA expansions. To date, more than a dozen such disorders have been found, and all of them affect the nervous system. Analysis of the rare families in which such diseases are common has revealed that expansion of the triplet repeats is linked to something called genetic anticipation, when a disease's symptoms appear earlier and more severely in each successive generation.

Battle of the Sexes

A process called imprinting, which occurs naturally in our cells, provides another example of how epigenetics affects gene activity. With most genes, the two copies work exactly the same way. For some mammalian genes, however, only the mother's or the father's copy is switched on regardless of the child's gender. This is because the genes are chemically marked, or imprinted, during the process that generates eggs and sperm.

As a result, the embryo that emerges from the joining of egg and sperm can tell whether a gene copy came from Mom or Dad, so it knows which copy of the gene to shut off. One example of an imprinted gene is insulin-like growth factor 2 (Igf2), a gene that helps a mammalian fetus grow. In this case, only the father's copy of Igf2 is expressed, and the mother's copy remains silent (is not expressed) throughout the life of the offspring.

Scientists have discovered that this selective silencing of Igf2 and many other imprinted genes occurs in all placental mammals (all except the platypus, echidna, and marsupials) examined so far, but not in birds. Why would nature tolerate a process that puts an organism at risk because only one of two copies of a gene is working? The likely reason, many researchers believe, is that mothers and fathers have competing interests, and the battle- field is DNA!

The scenario goes like this: It is in a father's interest for his embryos to get bigger faster, because that will improve his offspring's chances of survival after birth. The better an individual's chance of surviving infancy, the better its chance of becoming an adult, mating, and passing its genes on to the next generation. Of course mothers want strong babies, but unlike fathers, mothers provide physical resources to embryos during pregnancy. Over her lifetime, a female is likely to be pregnant several times, so she needs to divide her resources among a number of embryos in different pregnancies.

Researchers have discovered over 200 imprinted genes in mammals since the first one was identified in 1991. We now know that imprinting controls some of the genes that have an important role in regulating embryonic and fetal growth and allocating maternal resources. Not surprisingly, mutations in these genes cause serious growth disorders. Marisa Bartolomei of the University of Pennsylvania School of Medicine in Philadelphia is trying to figure out how Igf2 and other genes become imprinted and stay silent throughout the life of an individual. She has already identified sequences within genes that are essential for imprinting. Bartolomei and other researchers have shown that these sequences, called insulators, serve as "landing sites" for a protein that keeps the imprinted gene from being transcribed.

Starting at the End

When we think of DNA, we think of genes. However, some DNA sequences are different: They don't encode RNAs or proteins. Introns. Another example is telomeres—the ends of chromosomes. There are no genes in telomeres, but they serve an essential function. Like shoelaces without their tips, chromosomes without telomeres unravel and fray. And without telomeres, chromosomes stick to each other and cause cells to undergo harmful changes like dividing abnormally. Researchers know a good deal about telomeres, dating back to experiments performed in the 1970s by Elizabeth Blackburn, a basic researcher who was curious about some of the fundamental events that take place within cells. At the time, Blackburn, now at the University of California, San Francisco, was working with Joseph Gall at Yale University. For her experimental system, she chose a single-celled, pond-dwelling organism named *Tetrahymena.* These tiny, pear-shaped creatures are covered with hairlike cilia that they use to propel themselves through the water as they devour bacteria and fungi.

Tetrahymena was a good organism for Blackburn's experiments because it has a large number of chromosomes—which means it has a lot of telomeres! Her research was also perfectly timed, because methods for sequencing DNA were just being developed. Blackburn found that *Tetrahymena's* telomeres had an unusual nucleotide sequence: TTGGGG, repeated about 50 times per telomere. Since then, scientists have discovered that the telomeres of almost all organisms have repeated sequences of DNA with lots of Ts and Gs. In human and mouse telomeres, for example, the repeated sequence is TTAGGG.

The number of telomere repeats varies enormously, not just from organism to organism but in different cells of the same organism and even within a single cell over time. Blackburn reasoned that the repeat number might vary if cells had an enzyme that added copies of the repeated sequence to the telomeres of some but not all chromosomes. With her then-graduate student Carol Greider, now at Johns Hopkins University, Blackburn hunted for the enzyme. The team found it and Greider named it telomerase.

The telomerase enzyme, it turns out, consisted of a protein and an RNA component, which the enzyme uses as a template for copying the repeated DNA sequence. What is the natural function of telomerase? As cells divide again and again, their telomeres get shorter.Most normal cells stop dividing when their telomeres wear down to a certain point, and eventually the cells die. Telomerase can counteract the shortening. By adding DNA to telomeres, telomerase rebuilds the telomere and resets the cell's molecular clock.

The discovery of telomerase triggered new ideas and literally thousands of new studies. Many researchers thought that the enzyme might play important roles in cancer and aging. Researchers were hoping to find ways to turn telomerase on so that cells would continue to divide (to grow extra cells for burn patients, for example), or *off* so that cells would stop dividing (to stop cancer, for instance). So far, they have been unsuccessful. Although it is clear that telomerase and cellular aging are related, researchers do not know whether telomerase plays a role in the normal cellular aging process or in diseases like cancer.

Recently, however, Blackburn and a team of other scientists discovered that chronic stress and the perception that life is stressful affect telomere length and telomerase activity in the cells of healthy women. Blackburn and her coworkers are currently conducting a long-term, follow up study to confirm these intriguing results.

The other Human Genome

Before you think everything's been said about DNA, there's one little thing we didn't mention: Some of the DNA in every cell is quite different from the DNA that we've been talking about up to this point. This special DNA isn't in chromosomes— it isn't even inside the cell's nucleus where all the chromosomes are! So where is this special DNA? It's inside mitochondria, the organelles in our cells that produce the energy-rich molecule adenosine triphosphate, or ATP. Mendel knew nothing of mitochondria, since they weren't discovered until late in the 19th century. And it wasn't until the 1960s that researchers discovered the mitochondrial genome, which is circular like the genomes of bacteria.

In human cells, mitochondrial DNA makes up less than 1 percent of the total DNA in each of our cells. The mitochondrial genome is very small—containing only about three dozen genes. These encode a few of the proteins that are in the mitochondrion, plus a set of ribosomal RNAs used for synthesizing proteins for the organelle. Mitochondria need many more proteins though, and most of these are encoded by genes in the nucleus. Thus, the energy-producing capabilities of human mitochondria—a vital part of any cell's everyday health—depend on coordinated teamwork among hundreds of genes in two cellular neighborhoods: the nucleus and the mitochondrion.

Mitochondrial DNA gets transcribed and the RNA is translated by enzymes that are very different from those that perform this job for genes in our chromosomes. Mitochondrial enzymes look and act much more like those from bacteria, which is not surprising because mitochondria are thought to have descended from free-living bacteria that were engulfed by another cell over a billion years ago. Scientists have linked mitochondrial DNA defects with a wide range of age-related diseases including neurodegenerative disorders, some forms of heart disease, diabetes, and various cancers. It is still unclear, though, whether damaged mitochondria are a symptom or a cause of these health conditions.

Scientists have studied mitochondrial DNA for another reason: to understand the history of the human race. Unlike our chromosomal DNA, which we inherit from both parents, we get all of our mitochondrial DNA from our mothers. Thus, it is possible to deduce who our maternal ancestors were by tracking the inheritance of mutations in mitochondrial DNA. For reasons that are still not well understood, mutations accumulate in mitochondrial DNA very slowly compared to chromosomal DNA. So, it's possible to trace your maternal ancestry way back beyond any relatives you may know by name—all the way back to "African Eve," the ancestor of us all!

The Tools of Genetics: Recombinant DNA and Cloning

To splice a human gene (in this case, the one for insulin) into a plasmid, scientists take the plasmid out of an *E. coli* bacterium, cut the plasmid with a restriction enzyme, and splice in insulin-making human DNA. The resulting hybrid plasmid can be inserted into another *E. coli* bacterium, where it multiplies along with the bacterium. There, it can produce large quantities of insulin. In the early 1970s, scientists discovered that they could change an organism's genetic traits by putting genetic material from another organism into its cells. This discovery, which caused quite a stir, paved the way for many extraordinary accomplishments in medical research that have occurred over the past 35 years. How do scientists move genes from one organism to another? The cutting and pasting gets done with chemical scissors: enzymes, to be specific. Take insulin, for example. Let's say a scientist wants to make large quantities of this protein to treat diabetes. She decides to transfer the human gene for insulin into a bacterium, *Escherichia coli*, or *E. coli*, which is commonly used for genetic research. That's because E. coli reproduces really fast, so after one bacterium gets the human insulin gene, it doesn't take much time to grow millions of bacteria that contain the gene.

The first step is to cut the insulin gene out of a copied, or "cloned," version of the human DNA using a special bacterial enzyme from bacteria called a restriction endonuclease. (The normal role of these enzymes in bacteria is to chew up the DNA of viruses and other invaders.) Each restriction enzyme recognizes and cuts at a different nucleotide sequence, so it's possible to be very precise about DNA cutting by selecting one of several hundred of these enzymes that cuts at the desired sequence. Most restriction endonucleases make slightly staggered incisions, resulting in "sticky ends," out of which one strand protrudes.

The next step in this example is to splice, or paste, the human insulin gene into a circle of bacterial DNA called a plasmid. Attaching the cut ends together is done with a different enzyme (obtained from a virus), called DNA ligase. The sticky ends join back together kind of like jigsaw puzzle pieces. The result: a cut-and-pasted mixture of human and bacterial DNA. The last step is putting the new, recombinant DNA back into *E. coli* and letting the bacteria reproduce in a petri dish. Now, the scientist has a great tool: a version of *E. coli* that produces lots of human insulin that can be used for treating people with diabetes. So, what is cloning? Strictly speaking, it's making many copies of a gene—in the example above, *E. coli* is doing the cloning. However, the term cloning is more generally used to refer to the entire process of isolating and manipulating a gene. Dolly the cloned sheep contained the identical genetic material of another sheep. Thus, researchers refer to Dolly as a clone.

Rules Governing RNA's Anatomy Revealed

"RNA is a very floppy molecule that often functions by binding to something else and then radically changing shape," said Al-Hashimi, who is the Robert L. Kuczkowski Professor of Chemistry and a professor of biophysics. These shape changes, in turn, trigger other processes or cascades of events, such as turning specific genes on or off. Because of the RNA molecule's mercurial nature, "you can't really define it as having a single structure," Al-Hashimi said. "It has many possible orientations, and different orientations are stabilized under different conditions, such as the presence of particular drug molecules." A major goal in structural biology and biophysics is to be able to predict not only the complex three-dimensional shapes that RNA assumes (which are dictated by the order of its nucleic acid building blocks), but also the various shapes RNA takes on after binding to other molecules such as proteins and small-molecule drugs. Further, researchers would like to be able to manipulate the 3-D structure and resulting activity of RNA by tweaking the drug molecules with which it interacts. But to do that, they need to understand the rules that govern the anatomy of RNA.

The quest has parallels to the study of human anatomy, Al-Hashimi said. "Your body has a specific shape that changes predictably when you are walking or when you are catching a ball; we want to be able to understand these anatomical rules in RNA." Manipulating RNA is a much sought-after goal, given the recent explosion in vital cellular roles ascribed to RNA and the growing number of diseases that are linked to RNA malfunction. RNA performs many of its roles by serving as a switch that changes shape in response to cellular signals, prompting appropriate reactions in response. The versatile molecule also is essential to retroviruses such as HIV, which have no DNA and instead rely on RNA to both transport and execute genetic instructions for everything the virus needs to invade and hijack its host.

In earlier work, Al-Hashimi's team determined that rather than changing shape in response to encounters with drug molecules, RNA goes through a predictable course of shape changes on its own. Drug molecules simply "wait for" the right shape and attach to RNA when the RNA assumes the particular drug's preferred orientation, Al Hashimi said. But what rules control the predictable path of shapes the RNA molecule assumes? And are those rules the same for all sorts of RNA molecules? In the current work, Al-Hashimi's team investigated those questions.

"RNA is very similar to the human body in its construction, in that it's made up of limbs that are connected at joints," Al-Hashimi said. The limbs are the familiar, ladder-like double helix structures, and the joints are flexible junctions. The prevailing view was that interactions among loopy structures at the tips of the limbs played a role in defining the molecule's overall 3-D shape, much as a handshake defines the orientation of two arms, but Al-Hashimi's group decided to look at things from a different perspective. "We wondered if the junctions themselves might provide the definition," Al-Hashimi said. "If you look at your arm, you'll notice that you can't move it, relative to your shoulder, in just any way; it's confined to a certain pathway because of the joint's geometry. We wondered if the same thing might be true of RNA." To investigate that possibility, the researchers turned to a database of RNA structures and found that all structures with two helices linked by a particular type of junction called a trinucleotide bulge fell along the same pathway.

The team then went on to explore structures of RNA molecules with other kinds of junctions. All were confined to similar pathways, but the precise pathway of a given RNA depended upon structural features of its junction. Just as anatomical features of our shoulders, elbows, hips and knees define the range of motion of our arms and legs, the anatomy of RNA's junctions dictates the motion of its helices. Next, Al-Hashimi and coworkers wanted to understand how drug molecules cause RNA molecules to freeze in specific positions. In earlier work with an RNA molecule known as TAR, which is critical for replication of HIV and thus a key target for anti-HIV drugs, the researchers had found that certain drug molecules froze the RNA molecule in a nearly straight position, while others trapped the molecule in a bent conformation and still others captured positions between the two extremes. But because that project involved a wide variety of drug molecules, it was hard to figure out why certain ones preferred certain orientations.

To explore the issue more methodically, Al-Hashimi's group used a series of aminoglycosides (antibiotics that are known to target RNA) that systematically differed from one another in charge, size and other chemical properties. Size turned out to be the key: bigger aminoglycosides froze RNA in more bent positions; smaller ones favoured straighter RNA structures. Looking more closely, the researchers discovered that the aminoglycoside molecule nestles between two helices and acts like wedge, forcing the helices apart. Examination of other RNA structures bound to small molecules revealed that this rule is not specific to TAR but a more general feature of RNA-small molecule interactions.

"With these findings, it now should be possible to predict gross features of RNA 3-D shapes based only on their secondary structure, which is far easier to determine than is 3-D structure," Al-Hashimi said. "This will make it possible to gain insights into the 3-D shapes of RNA structures that are too large or complicated to be visualized by experimental techniques such as X-ray crystallography and NMR spectroscopy. The anatomical rules also provide a blueprint for rationally manipulating the structure and thus the activity of RNA, using small molecules in drug design efforts and also for engineering RNA sensors that change structure in user-prescribed ways."

Discovery of the Hybrid Helix and the First DNA-RNA Hybridization

Here we describe early research on RNA structure and the discovery of the DNA-RNA hybrid helix, a key component of information transfer. More than 50 years ago it was realized that the interaction between DNA and RNA was at the core of molecular biology. The problem was chemical in nature: could two different types of molecules interact and serve in the transmission of biological information?

In the 1950s it was widely assumed that "DNA makes RNA, RNA makes protein." This was not based on experimental evidence that DNA and RNA could combine but was more in the nature of an intuitive belief. However, by early 1960 I was finally able to carry out a direct experiment, the first DNA-RNA hybridization. In 1960, messenger RNA was still 1 year in the future, and there was not a great deal of understanding of the major components of information transfer. The DNA double helix proposed by Watson and Crick in 1953 clearly suggested that information was contained in the order of nucleotides, but during the 1950s our understanding of RNA was fragmentary.

The origins of the 1960 experiment go back to the mid-1950s. In 1954, while at Caltech, Jim Watson and I had been trying to find out if RNA by itself could form a double helix, but the fibre diffraction studies of RNA gave inconclusive results. The fuzzy diffraction patterns all looked alike, but unlike DNA, the RNA base ratios were all different in the samples we examined. It was clear that RNA was more complex than DNA, and its structure was unknown. In their 1953 paper Watson and Crick pointed out that it was probably impossible to form their double helix with ribose due to van der Waals interference of the 22-OH with the structure. Thus, it was likely to be different.

A New Type of Polymer Chemistry

The research changed dramatically at the National Institutes of Health (NIH) in 1956 when David Davies and I began working with synthetic polyribonucleotides made using the polynucleotide phosphorylase enzyme discovered by Grunberg-Manago and Ochoa. When we mixed together polyriboadenylic acid (poly(rA)) and polyribouridylic acid (poly(rU)) and pulled fibres, they yielded a clear diffraction pattern of a double helix. The two molecules had combined to make an RNA double helix! This was a new type of chemical reaction, one that involved thousands of units binding specifically to each other in an extended array.

Furthermore, no polymer reaction had ever been seen in which the monomers from two polymers bound together with great specificity. Two weeks after sending off the 1956 *Journal of the American Chemical Society* note, I wrote a letter to my postdoctoral mentor, Linus Pauling, describing these results. The letter reveals a sense of incredulity on my part that this reaction could happen and that it was "completely reproducible." The experimental demonstration that these molecules could form a double helix seems obvious today. However, it was a considerable surprise at that time and was often greeted with skepticism. Most biochemists felt that a double helix could only be made by an enzyme, such as the one discovered by Arthur Kornberg and associates, which appeared to replicate the DNA double helix. Polymer chemists felt that very long molecules involving thousands of nucleotides would probably become entangled and could not sort themselves out to form a regular double helix. Still other researchers, on theoretical grounds, felt that two highly negatively charged polymers were unlikely to combine to make a single structure.

However many scientists were receptive. In the early fall of 1956, a McCollum Pratt meeting was organized at Johns Hopkins University in Baltimore around the subject of the "Chemical Basis of Heredity." It was an excellent meeting with all of the major research workers in the field. In my talk I included a discussion of the specificity of the interaction and the fact that these long molecules seek each other out in solution and adopt this elongated helical form. Julian Huxley, a prominent scientist and writer, came up to me after my talk and warmly congratulated me for having discovered "molecular sex." Gradually, the idea became fixed in the thinking of biologists and biochemists that these long nucleic acid molecules had considerable flexibility and could be made to form helical structures in solution. Although no one used the phrase at the time, this was the first hybridization reaction, and it represented a paradigm shift in the way chemists and biochemists thought about macromolecular nucleic acids.

Two Different Structures

Watson and Crick pointed out that RNA would not form the DNA structure. This was confirmed as the RNA x-ray diffraction pattern revealed significant differences from the DNA double helix pattern. The diameter of the RNA double helix was 6 Å larger than the DNA double helix. Analysis of the diffraction pattern revealed that the first layer line of the RNA-RNA duplex was stronger than the second—a reversal from that seen in the diffraction pattern of the DNA-DNA duplex. Another significant difference was that alterations in relative humidity changed the DNA double helix, producing both the A and B forms, whereas the RNA double helix seemed invariant to changes in relative humidity.

The reaction between poly(rA) and poly(rU) was associated with a drop in optical density in the ultraviolet, a property that could be used to identify the 1:1 stoichiometry and analyse the reaction quantitatively. Only much later was it realized that the RNA duplex structure was close to the dehydrated A conformation of DNA duplex. However, by 1960, it was clear that DNA and RNA duplexes were significantly different. In 1957, together with Felsenfeld, we discovered that the poly(rA)·poly(rU) duplex could take on a third strand of poly(rU) to form a triple helix. We pointed out that this could be associated with forming two hydrogen bonds between the incoming uracil O4 and N3 and adenine N6 and N7, an interaction of bases that was confirmed 2 years later by Hoogsteen in a single crystal x-ray analysis. Between 1956 and 1960 a number of experiments were carried out with polyribonucleotides of different composition, and their interactions could generally be explained in terms of the ability of the bases to form at least two hydrogen bonds in the centre of the molecule.

Could DNA and RNA interact? Polymers of ribonucleotides were available because of the discovery of polynucleotide phosphorylase. This made it possible to produce polyribonucleotides with a variety of compositions, but there was no analogous method of producing DNA polymers. The question "How does DNA make RNA?" remained an open issue. Several biochemists were trying to isolate the enzyme known as RNA polymerase which is dependent upon a DNA template.

Experiments by Stevens, Hurwitz, and Weiss developed preparations that had some activity, but they were not purified enough to show what was actually happening. In a reflective article published in 1959 called "An Analysis of the Relation between DNA and RNA," I surveyed the various possibilities for DNA-RNA interactions. In particular, I asked whether RNA synthesis could be based on a double-stranded primer or a single-stranded primer of DNA. The discovery of several triple helical complexes of RNA molecules made it reasonable to consider a model in which RNA nucleotides were assembled by binding in a sequence-specific manner to double-stranded DNA. The analysis showed that such a model was unlikely because there were not enough stereospecific interactions to specify an RNA molecule. I concluded that it was likely to be based on forming an RNA strand on a single-stranded primer. These conclusions were fortified by the recent discoveries that denatured DNA could provide a primer for Kornberg's DNA polymerase enzyme, and furthermore, Sinsheimer had discovered that the virus ÕX174 had a single-stranded genome and it was a primer for the DNA polymerase. This reinforced the idea that single strands were adequate templates. In the same article, I speculated that RNA was likely to be the first polynucleotide molecule in the origin and early evolution of life and also pointed out the possibility of an enzyme that used a single-stranded RNA primer to make a DNA strand; it was called reverse transcriptase when it was discovered 10 years later.

Formation of a Hybrid Helix

Could a DNA-RNA hybrid helix form and have the stability needed so that it could be used for information transfer in view of the different physical properties and geometries of the RNA and the DNA duplex? I was finally able to address this problem with the chemical synthesis of oligodeoxythymidylic acid by Khorana and colleagues. Khorana kindly gave me a reaction mixture, which I fractionated and took the longer molecules to see if they would react with polyriboadenylic acid. The experiment worked. The discovery in 1960 that these could form a double helix represented the first experimental demonstration that a hybrid helix could be a method for the transfer of information from DNA to RNA. The evidence was based on measurements of the hypochromism, which occurs when helical molecules are formed, and the changes in sedimentation rate associated with complex formation. These results indicated that DNA could make RNA by using a single-stranded template to make a complementary RNA strand, and it was a model for RNA polymerase activity. However, the discovery of messenger RNA was still 1 year in the future.

It is interesting that 1 year later in 1961, with a more highly purified preparation of RNA polymerase and using as a template the same oligodeoxythymidylate synthesized by Khorana, Hurwitz was able to synthesize polyriboadenylic acid using his enzyme preparation. This proved that a single-stranded template was adequate for RNA polymerase.

The publication in the summer of 1960 was the first demonstration of DNA-RNA hybridization although that term had not yet been invented. That particular hybridization is still widely used today in that immobilized oligo(dT) molecules are used to isolate eukaryotic messenger RNA through their poly(rA) tails. At the same time in 1960, Marmur, Doty, and colleagues found they could take denatured DNA molecules and hold them for a prolonged period at an intermediate temperature, called an annealing temperature, thereby allowing the DNA molecules to find each other and re-form a double helix. After presenting my work on DNA-RNA interactions at the Gordon Conference in 1960, Sol Spiegelman and Ben Hall came up to me and said they were inspired to try that reaction with a viral system. One year later they combined my work with the annealing work of Marmur and Doty and found that a newly synthesized RNA strand from T2 virus infection could be similarly annealed with DNA from the virus to make a hybrid helix. Thus, by 1961 both DNA-DNA and RNA-DNA hybridizations were available for a variety of biological studies.

The significance of the discovery of the DNA-RNA hybrid helix is relevant not only to understanding the activity of RNA polymerase, but it also explains how reverse transcriptase works, as well as telomerase, retrotransposons, and a variety of other interactions in which DNA and RNA strands combine. Today, fluorescence *in situ* hybridization is used in which DNA-RNA hybrids are formed to identify specific areas of the genome. Similarly, microarray gene expression profiling studies use DNA-RNA hybrid formation and are dependent on this 1960 discovery.

The full structural analysis of how these RNA and DNA chains accommodated each other was not revealed until much later. In 1982 the first DNA-RNA single-crystal hybrid molecule was solved in my laboratory. It showed that a DNA decamer containing four ribonucleotides formed a hybrid segment that stabilized the entire molecule in the A conformation. In 1992, we carried out a single-crystal x-ray analysis of an Okazaki fragment in which the nucleating ribonucleotides that initiate DNA synthesis in DNA replication were crystallized on DNA. Again, the presence of a few ribonucleotides was sufficient to convert the entire fragment into the A conformation.

The central role of the ribose ring puckers that differ in DNA and RNA is widely understood today. The van der Waals crowding of the ribose C22 oxygen determines the RNA pucker in the RNA double helical A conformation. This interaction provides a sufficient energy barrier to prevent a change in pucker in double helical RNA, in contrast to the facile manner with which the deoxyribose in DNA changes its ring pucker on lowering the water content to form A-DNA or to accommodate the presence of RNA in a hybrid double helix. The hybrid DNA-RNA helix remains the bedrock of information transfer in biological systems. Indeed, the existence of a hybrid helix seems so obvious today that young research workers simply take it for granted. There is little realization of the extent to which, almost a half-century ago, scientists wrestled with problems of understanding how different polymers can react together to make a stable structure. The roots of our understanding of hybrid helix formation go back to 1960 and even further to the mid-1950s.

Genetic Connections Between Organisms

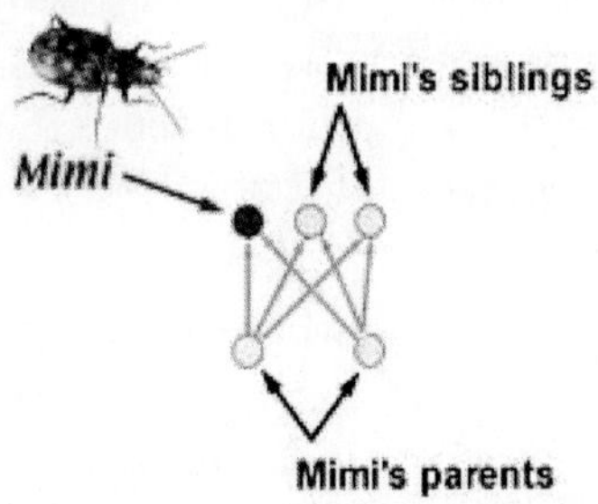

Figure: *A diagram of Mimi's immediate family. The passage of genes from parents to offspring is indicated by the green lines.*

What do we Mean when We Say all Organisms are Genetically Related?

We are all familiar with the genetic relationships between parents and their offspring. Kids usually resemble their parents and their siblings, because offspring inherit genes from their parents. But how do you get from these obvious, close genetic relationships to the more distant genetic relationships between species that are as different as daisies and elephants?

Consider an Individual Organism, a Beetle Called Mimi.

Like many other organisms, the beetles in Mimi's species reproduce sexually. This means that a parent passes on copies of half of his or her genes to each offspring. An offspring's genome then represents a combination of parental genes. Mimi's family lives under a log, with a group of other beetles. New beetles are born every summer, and they live for about one year. So each year there is a new generation of beetles. As you can see in Figure 2, Mimi does not only have two full siblings, but also two half siblings, since one of her parents produced offspring with another beetle.

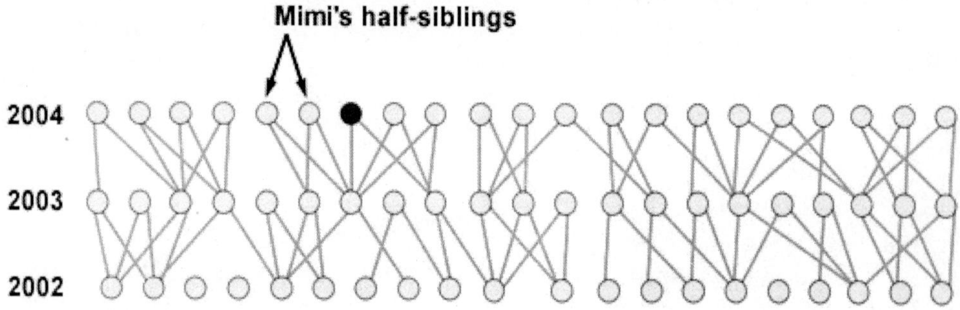

Figure: *The genetic relationships for the last three generations of beetles under Mimi's log.*

Mimi's log is on a beach where there are a number of other logs housing groups of beetles. Members of Mimi's group sometimes meet beetles from these other logs and produce offspring with them. Occasionally, a beetle from another beach may fly in and participate in the production of offspring. Similarly, beetles from Mimi's beach may visit other beaches and mate with beetles there. Thus, Mimi is part of a larger population of beetles, extending over several beaches.

The genetic connections among organisms within a sexually reproducing populations form a tangled web. Through generations, the genes of an individual are passed along a chain of offspring, i.e., children, grandchildren, great-grandchildren, and so on.

Through the process of interbreeding, the individual's genes are combined at each step with the genes of other individuals. However, these genetic connections do not extend much between different species. Organisms generally look for mates within their own species, and so each species forms a reproductive community. As a result, each species descends through time as a bundle of genetic connections isolated from other such bundles formed by other species.

Occasionally, a species might be broken into two isolated reproductive communities. This could happen, for example, if some geographic barrier arose (e.g., a river broadening, sea levels rising, a mountain range lifting) that would prevent beetles from southern beaches from moving to northern beaches and vice versa. If such a barrier remained in place for a long time, the two isolated reproductive communities would evolve into separate species, and each would become its own isolated bundle of genetic connections. Something like this must have happened a long time ago when Mimi's species, *Bembidion zephyrum*, evolved from the ancestor it shares with its closest relative, *Bembidion levettei*.

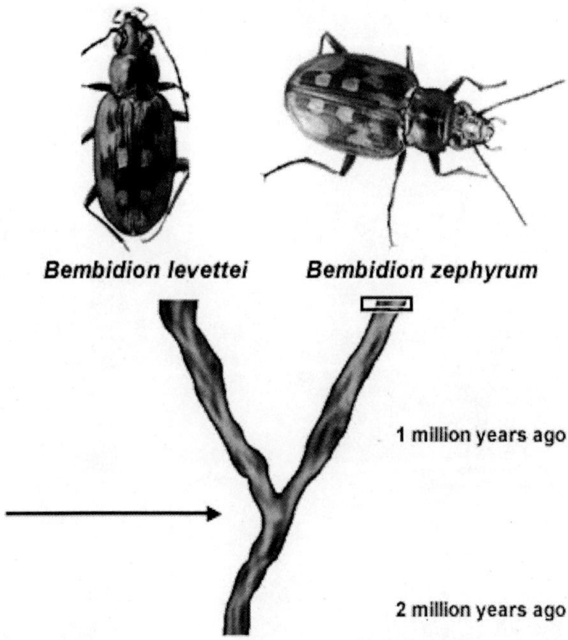

Figure: *Genetic relationships between two closely related species,* Bembidion levettei *and* Bembidion zephyrum. *Recent generations of beetles in Mimi's species are indicated with the blue frame. The arrow points to the common ancestor of the two beetle species.*

If we went further back in the evolutionary history of Mimi's genetic lineage, we would see a sequence of beach beetle lineages splitting in two and giving rise to new species. Once the genetic ties between two lineages are severed, they are likely to diverge in their looks and habits, as ancestral features are modified independently in each species. So each species will be characterized by a unique set of traits, and species that are more closely related will generally be more similar to one another.

During the evolutionary history of beach beetles, some species would have gone extinct, while others, like Mimi's *Bembidion zephyrum*, would have flourished, producing a new generation of beetles each year. This process has resulted in an ever growing phylogenetic tree of species lineages, with some branches pruned in the past, and the species living today sitting like leaves at the tips of actively growing branches (Figure). Over the course of hundreds of millions of years, the splitting and subsequent divergence of lineages has produced the Tree of Life, which has as its leaves the many species of organisms we see around us today.

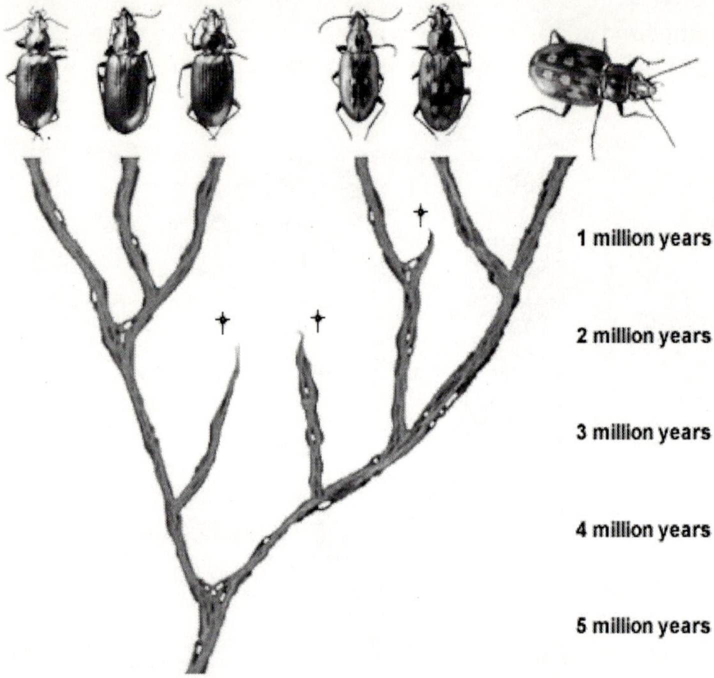

Figure: A phylogenetic tree of beach beetles. Some branches have gone extinct in the past, while others represent species living today.

'Tree of Life' has lost a Branch, According to Largest Genetic Comparison of Higher Life Forms Ever

The discovery by Norwegian and Swiss researchers has gained attention from biologists worldwide. The findings come from the largest ever genetic comparison of higher life forms on the planet. Of 5000 genes examined, researchers identified 123 common genes from all known groups of organisms; these common genes have been studied more closely.

Lost a Branch

"The results were pretty astounding. All non-bacterial life on Earth—called eukaryotic life— can now be divided into four main groups instead of the five groups that we have been working with up to now," says Kamran Shalchian-Tabrizi, an associate professor from the University of Oslo's Department of Biology who has also worked with the Department of Zoology and Animal Biology and the Department of Genetic Medicine and Development, at the University of Geneva, Switzerland. The Tree of Life (see illustration) has, through the discovery that the two formerly separated branches share a similar evolutionary history, lost one of its branches, and this will both improve and simplify quite a bit of scientific work in the future.

Important Discovery

"Kinship says a lot about shared traits. Our findings can be important in many fields, such as in the study of the development of life and in the manufacture of new medicines" says Shalchian-Tabrizi in an interview with the University of Oslo's research magazine Apollon. "Our knowledge of organisms and the development of medicines are often based on comparative studies across species. It is, therefore, essential that we know the relationships between the largest groups in the great diversity of eukaryotes," he adds.

The research group has, for example, found that brown algae and silica algae, and groups of single cell organisms like the malaria parasite, marine foraminifera, and the green sun animalcule (acanthocystis turfacea) actually belong to the same group. Previously, these species were thought to be completely unrelated.

"The work that we published in the August edition of PLoS One means that the description of the Tree of Life must be revised in new textbooks," says Professor Kjetill S. Jakobsen from the University of Oslo's Centre for Ecological and Evolutionary Synthesis (CEES). He is also a member of the Microbial Evolution Research Group (MERG), led by Shalchian-Tabrizi, at the Department of Biology. MERG is one of 16 groups that the Faculty of Mathematics and Natural Sciences believes may have the potential to develop into new Centres of Excellence.

The New Branch

All life on Earth can be divided into two essentially different life forms—eukaryotes and prokaryotes. The eukaryotes gather their genetic material in a nucleus, while the prokaryotes (bacteria and archaea) have their genetic material floating freely in the cell. Eukaryotic organisms—such as humans—can, as a result of the new findings, be divided into the following four categories:

"The SAR group has to some extent been identified earlier, but we could not know if it was a correct observation because we lacked statistical data. To get that data, we first had to reconstruct the entire eukaryote tree with the help of these 123 genes. Chromalveolates and rhizaria were clearly separate groups until we published our results," says Shalchian-Tabrizi.

"To make the picture a little less clear, one branch of chromalveolates is still in no man's land. It may be that these also belong to SAR, but we will require additional genes and genomes to study this. We have set our sights on doing that in the course of the next few years," he adds.

Fewer Events

"The Tree of Life tells the story of life on Earth, and our research can say something about how quickly life developed. Our discovery suggests that there were fewer big "events" than we have previously assumed in the development of higher life forms. The more we know about the branches on the Tree of Life, the more we can find out about life's Big Bang, the beginning of life on Earth," says Shalchian-Tabrizi.

Three billion years ago, there was only bacteria and Archaea. Eukaryotic life, which comprises all multi-celled organisms, developed in the sea—probably between 1.2 and 1.6 billion years ago. It was not before about 500 million years ago that the first creatures crept onto land.

"By digging down into the historical layers with the help of phylogenetic reconstruction, where we can find out about kinship between organisms at the genetic level and we can find answers to questions about how new traits developed. We are working, in a matter of speaking, with genetic archaeology. In this manner, we can also discover the cause of the Earth's biological diversity," says Jakobsen.

10

Characteristics of Genetic System

X-Chromosome as Type

The chromosome combinations present in the fertilized egg, and this has brought to light some of the methods by which the chromosomes act on development and characteristics. We now Examine other effects of altering chromosomes.

We shall have to examine the effects of altering all the different classes of chromosomesýÿthe X-chromosome, the Y-chromosome, and the different autosomes. Each of these three groups (X, Y, and autosomes) gives rise, as we shall see, to a different type of inheritance. The simplest relations, from the experimental point of view, are presented by the X-chromosomes. We therefore deal first with these.

Effects on Altering the X-Chromosomes

The X-chromosomes take such a course in passing from generation to generation that it is possible to follow the descendants of a particular Xchromosome (that present in the original male parent, for example), knowing in which individuals they are present, in which they are absent. In certain individuals (males, in organisms of Group I) there is but a single X-chromosome instead of a pair.

These relations make it a relatively simple matter to discover the distinctive effects of a particular X-chromosome. The effects of X-chromosomes will therefore be dealt with somewhat fully as a type of chromosomal action. Later the action of the other chromosomes will be taken up.

Abnormal Distribution of X-Chromosomes

What will happen if the X-chromosomes by accident become irregularly distributed? We know that sometimes the X-chromosomes are indeed irregularly distributed. Will the sex-linked characteristics continue to follow them; will the defective characteristics show the same irregular distribution as do the X's? Such cases in great number have been fully observed. It is found that the sex-linked characteristics do indeed follow the X-chromosomes wherever they go. This proves conclusively (if there were any possible doubt in view of the extraordinary course normally followed by such characters) that it is indeed the X-chromosomes on which the characters depend. The matter is one of importance and interest, so that it will be worth while to examine carefully certain typical cases of the result of irregular distribution of X-chromosomes.

Normal Distributions

Sometimes the body of Drosophila is yellow instead of the normal grey; this is a recessive character due to a defect in the X-chromosomes. Suppose that we mate together a female that has a yellow body and a male that has the normal grey programmes. The female has two modified X-chromosomes, which we may represent by XX. The male has its X unmodified; we may represent its cells as XY. Normally in forming the germ cells, the two X's of the female separate into different germ cells, and all the ova receive one of the modified X's. The normal male produces germ cells of two classes, X and Y. When the germ cell X from the female unites with Y from the male, sons are produced, with the constitution XY; while X from the female with X from the father gives daughters XX. Since the yellow programmes dependent on X is recessive, the daughters XX have the normal grey programmes, like the father.

But the sons XY have only one X, and as this is the defective one, from the mother, the sons are yellow, like the mother. The normal result of such a mating therefore is that all the sons are recessive like the mother, all the daughters dominant, normal, like the father. This is what occurs when the X-chromosomes are distributed to the germ cells in the usual way.

Abnormal Distribution

In forming germ cells both X's of the female go together to a single ovum, while other ova receive no X. In a race discovered by L. V. Morgan the two X-chromosomes in the female were partly united, so that they almost invariably thus go together to one ovum, leaving other ova without an X. In this race the X-chromosomes were so modified as to produce the recessive yellow programmes described in the preceding paragraph. We therefore have an opportunity to determine the course of inheritance of the yellow and the grey programmes when the X-chromosomes are thus abnormally distributed. Represent as before the X-chromosomes that produce the yellow programmes by X. Thus one set of ova receive XX, while another set receive no X. When the germ cells XX united with the germ cells Y from the normal father, there were produced daughters XXY, in which both X's came from the mother, and these daughters were yellow like the mother (since no normal X was present). Sons were produced by the union of ova that contained no X with sperms that carried X; this gave sons XO, with their single X from the normal father, instead of from the mother, as is usually the case. And such sons had the normal grey body programmes of the father, instead of the yellow body programmes of the mother, as happens normally.

Thus with a change in the distribution of the X's there is a corresponding change in the method of inheritance. If the sons receive their X from the recessive mother, as in the normal cases, they are recessive like the mother. But when, through non-disjunction, the sons receive a dominant X from the father, they are dominant like the father. Similarly, in the normal case the daughters receive a dominant X from the father and are therefore dominant like the father. But in cases of non-disjunction the daughters receive both their JPs from the recessive mother, and are therefore recessive like the mother. In sum, when the X's are normally distributed, such matings give 'criss-cross inheritance'; sons like the mother, daughters like the father. But when the X's are not so distributed, there is no criss-cross inheritance; sons are like the father, daughters like the mother. The dominant and recessive characters follow the respective X-chromosomes, whether these are distributed normally or abnormally.

Such experiments have been repeated many times, and with other sex-linked characters. Always the characteristics follow the distribution of the X-chromosomes, however these are distributed. It may be concluded with certainty that it is the presence of X-chromosomes of a modified type that causes the appearance of the particular sex-linked characters that are manifested. The rules of distribution of sex-linked characters are the rules of distribution of the X-chromosomes.

Dominance and Recessiveness in Sex-linked Characters

As we have seen, and shall see further, many defective conditions are due to defects in certain X-chromosomes. These follow from generation to generation the distribution of these X-chromosomes. Most of these bodily defects are manifested only in individuals in which the defective X- chromosomes are the only kind present; that is, these defects are recessive.

But it is important to observe that in such cases the normal condition of the organism likewise follows the distribution of certain X-chromosomes.

Haemophilia follows certain X- chromosomes. But in the same matings in which this occurs, the healthy condition of the blood follows certain other X-chromosomes. If a defective mother is mated with a normal father, the sons are defective because they receive only the mother's defective X-chromosomes. But in the same way the daughters are normal because they receive the father's normal X-chromosomes. In the normal individuals the X-chromosome plays a part, just as it does in the defective individuals. In the normal individuals it is healthy and supplies what is required for normal development, while in the defective individuals it fails to supply what is required.

In the case of most defects this normal condition is dominant over the recessive condition; that is, when both are present, the normal chromosome is the one that prevails in its effect on the individual. There are defective characters, however, in which the defective condition is dominant or partly dominant.

In these cases, when a defective X and a normal one are present together, the defective one produces its effect, as in the case of bar-eye, described earlier. In many such cases the defect is less marked when a normal chromosome is present as well as a defective one; then the defective condition is said to be partly dominant. Thus whenever an individual bearing defective X-chromosomes is mated with a normal individual, we have both dominant and recessive characters, the course of which may be followed in the later generations the dominant character being commonly the normal or usual condition. How this works out is shown in our next paragraphs.

Tests for Sex-linked Inheritance

By observing the distribution of the dominant and recessive characters among the offspring of certain matings, a test for determining whether given characters depend on the X-chromosomes is supplied. The test consists in making what are called reciprocal crosses.

- On the one hand mate a dominant female with a recessive male;
- Also mate a recessive female with a dominant male.

Represent a dominant X-chromosome by a capital X, a recessive X-chromosome by a lower-case letter x. In the group of organisms with which we have been dealing, the female has two X-chromosomes, the male but one. The daughters receive an X from each parent, the sons an X from the mother only. A dominant female will be represented by XX, a recessive male by xo; similarly a recessive female is xx, a dominant male is XO.

The two ma tings and their results will then be represented as follows:

- XX by xo gives X + XO Dom. Mother Rec. Father Dom. Daughters Dom. Sons
- xx by XO gives X + xo Rec. Mother Dom. Father Dom. Daughters Rec. Sons

Thus when the mother is dominant, the father recessive, all the children are dominant, like the mother. When the father is dominant, the mother recessive, the daughters are dominant like the father, the sons recessive like the mother; this is called 'criss-cross inheritance'. Whenever reciprocal crosses give these results, it is certain that the two diverse characteristics (dominant and recessive) result from differences in the X-chromosomes of the two parents. 'Criss-cross inheritance' is particularly useful in showing at once that we are dealing with sex-linked inheritance.

In every case where such results are produced, further tests show that the two characteristics follow in later generations the two different kinds of X-chromosomes wherever they go. It is mainly by the use of these tests that the many different characteristics dependent on diversities in X-chromosomes have been discovered.

Sex-linked Inheritance in Group II

Sex-linked inheritance as we have just described it was originally discovered in animals belonging to Group I, in which the females have two X-chromosomes, the males but one (with or without a Y- chromosome). But when the tests we have just described were applied to birds and to certain other animals, it was discovered that there is a second group (Group II) in which it is the male that has two X-chromosomes, the female but one. In the common fowl, reciprocal crosses were made between 'barred' fowls (Plymouth Rock) and black fowls (Langshan). Here the 'barred condition was found to be dominant.

The results were:

- Barred father by black mother=Barred sons + barred daughters.
- Black father by barred mother = Barred sons + black daughters.

Here the second mating gives 'criss-cross inheritance', showing that we are dealing with sex-linked characters dependent on differences between the X-chromosomes of the parents. But, in the first mating above, all the offspring are like the father instead of like the mother, which latter was the case in the matings. The father in this case has the dominant characteristic (barred), and all the offspring are dominant like the father instead of like the mother. When the mother is dominant ('barred') the result in the fowls is criss-cross inheritance; while, in the cases before described, criss-cross inheritance occurs when it is the father that is dominant. In fact, in the birds, the males and females simply exchange roles, as compared with their roles in the organisms of Group I.

Since the roles of males and females are interchanged in the two groups, their chromosomal conditions must also be interchanged; that is, since in Group I the female has two X's, and in Group II the male plays the same role as the female of Group I, the male of Group II must have two X's, the female but one. The chromosomal conditions in the two groups are shown in the above tabulation, X signifying dominant, while x signifies recessive, and o signifies the lack of X (whether Y is present or not).

Careful examination shows that this is the only way in which the results in Group II can be produced. Birds and certain moths show inheritance of the kind typical for Group II. In both these it has since been found under the microscope that the males have indeed one more chromosome than the females.

Role of Development in X-Chromosomes

A great many characteristics, of various kinds, have been found to be the result of modifications of particular X-chromosomes, and thus to follow in later generations the distribution of the descendants of those chromosomes. Examination of a number of these is desirable, both for their own importance and for the light they throw on the functions of the X-chromosomes.

In man, the following characteristics, among others, dependent on diverse types of X-chromosomes, are known from the fact that they show typical sex-linked inheritance.

- *Haemophilia*: Lack of coagulability in the blood. This results from a serious defect in certain X-chromosomes. The existence of defective X-chromosomes having this result shows that the normal X-chromosomes play a part in supplying something necessary for producing normal blood that coagulates properly.
- *Programmes Blindness*: The fact that defectiveness in X-chromosomes causes programmes blindness shows that the normal X's play a role in producing the normally functioning eyes.

Night Blindness

- Inability to see in a poor light.
- Near-Sightedness of certain types.

Progressive atrophy of the muscles. The fact that defects in X-chromosomes have this effect shows that the normal X's play a role in the normal functioning of the nerves and muscles.

A considerable number of other sex-linked characters are known in man. But man is a very unfavourable organism for the study of inheritance. Yet even the little that is known of sex-linked inheritance in man shows that the X-chromosome plays a role in many diverse bodily functions.

For a more complete idea of its role, some organism that can be bred experimentally must be examined. For this purpose the fruit-fly, Drosophila melanogaster, is the best organism to select, since it has been studied more extensively than any other. In the fruit-fly we find peculiarities of the following types that are dependent on alterations in particular X-chromosomes:

- *Many different eye colours*: The normal eye programmes in this organism is a certain shade of red. Defects in different X-chromosomes result in producing, in the individuals that bear these defective chromosomes, many different shades of red, varying from deep red to a very light red, and thence to 'buff', 'ivory, and 'white'. More than a dozen different types of eye programmes are known to result from modifications of the X-chromosome. It is clear that the normal X-chromosome plays an important role in producing the normal eye programmes. Structural peculiarities of the eye, such as 'bar-eye', 'facet eye', 'furrowed eye', and the like, are known to result from modifications of the normal X-chromosome.
- *Wing modifications*: Many different conditions of the wings are known that depend upon diversities among X- chromosomes borne by different individuals. These affect all sorts of features of the wings: size, form, venation, function. The normal X-chromosome obviously plays an important role in the full and normal development of the wings.
- *Body colours and markings*: Modifications of the X- chromosomes produce the body colours yellow, sable, tan, chrome, lemon, green, and the like, in place of the normal grey. Other changes in X alter the distribution of pigment on the body, giving the characters dot.
- *Body structure*: A defect in X produces irregularities in the abdomen, known as 'abnormal abdomen'.
- *Legs*: A defect in X-chromosomes causes abnormal development of the legs; some of them are wholly or partly reduplicated, so that the total number of legs is increased.
- *Bristles*: Modifications in X-chromosomes result in various different changes in the bristles that are scattered over the body.

Many physiological conditions and functions likewise depend upon the X-chromosomes, since they are changed when these chromosomes are modified. Among these are the following:

- *Positive reaction to light*: Drosophila with normal X-chromosomes fly towards a source of light. Those having defective X-chromosomes of the kind that produce a tan-coloured bodydo not fly towards a source of light.
- *Weakness and short life*: Most of the, defective X-chromosomes that cause structural or other changes in the body (different eye colours, wing forms, and the like) produce likewise weakness and short life. The individuals bearing them are less resistant to bad conditions, and live for a shorter time than the individuals that bear the normal X-chromosomes.
- *Life and death*: Some X-chromosomes have defects that are so severe that the individuals bearing them will not live and develop, unless there is present also a normal X-chromosome. Such defects are known as lethals. The presence of such lethal defects in the X-chromosome has the result that males bearing them die, since they have but one X; while females usually live, since they have an additional X that is often normal.

If a mother that has one lethal X and one normal one is mated to a father that has a normal X, the result is that half of the sons which receive the mother's defective X fail to develop, while the other half of the sons,receiving the mother's normal X, live and develop.

All the daughters live, since they all receive a normal X-chromosome from the father. The consequence is that in such families there are twice as many daughters as there are sons.

In families of 100 to 200, as occur in Drosophila, this is very striking. X-chromosomes are known to play in other organisms roles similar to those mentioned above for Drosophila, although in no other organism has the matter been so fully studied.

We may summarize what has been brought out above in the following statements:

- Many defects that appear in individuals are the result of defects in the X-chromosomes that they bear.
- Some of these defects are dominant; they are manifested in all individuals in which the defective X is present.
- But most such defects are recessive; they are manifested only in individuals in which the defective X is the only kind present.
- When both a normal X and a defective X are present, in most cases the normal X performs the required functions, so that the individual is not defective.
- Since the female has two X-chromosomes, while the male has but one, such defects are more frequently manifested in the males. Usually one of the X's present in the female is without the defect, so that she is not defective. But when the male carries a defective X-chromosome, the defect is manifested.
- It is thus advantageous, so far as the occurrence of defects is concerned, to have two X-chromosomes rather than one.
- This gives the female a considerable advantage over the male in these respects. There are many different types of defects due to defective X-chromosomes; almost all of them are more common in males than in females. Some of these defects seriously injure the health, or even result in death, if the defective X is the only kind that is present. It is well known that, in general, the death rate is higher in males than in females; this is probably due to the fact that males have but one X-chromosome.
- Among the different individuals of a species are scattered a great number of different types of X-chromosomes having different effects on development. Many of the diverse X-chromosomes are distinctly defective, causing personal defects in the individuals that bear them. Others are not defective.
- Defects or modifications in X-chromosomes affect in different cases all parts and functions of the organism.

The Role of the X-Chromosomes in Development

What do the facts brought out in the preceding sections show as to the physiological functions of the X-chromosomes; as to their role in development? For every kind of effect resulting from defects or modificacations in the X-chromosomes, there is a corresponding (opposed or diverse) action of normal or unmodified X-chromosomes. Since defective X-chromosomes of a certain type produce in Drosophila short, ill-developed wings, it follows that the normal X-chromosomes act on the development in such a way as to give long, well-developed wings.For if we substitute the normal X for the defective one, this causes the normal well-developed wings to be produced.

Similar reasoning applies to all the defective conditions that result from defective X's; the normal X's act in such a way as to produce the corresponding normal conditions. It follows therefore that the normal X-chromosomes play a role in the production of eye programmes and structure, body programmes and structure, in the development of the legs, the wings, the bristles, in producing normal health, resistance and vigour, and in various physiological processes. We know, further, that they play a most important part in determining sex, with all that this includes of structural and physiological influence.

Clearly therefore the X-chromosomes play a most important part in development. Their action is not limited to any single part of the body, nor to any single class of functions. They enter into the processes of development in such a way as to influence all parts of the body, all functions of the body. They begin to influence development very early, as was seen in considering the development of sex. And they play important roles in such relatively late processes as the production of eye colours.

It can hardly be doubted that they enter into the developmental activities from practically the beginning, influencing all the later processes that occur.

Autosomes and T-chromosomes

The X-chromosomes have been dealt with because of the great advantages they offer for experimental study. It was seen that they affect development in many ways, and that in a given species there are many different types of X-chromosomes in the different individuals, causing them to show different characteristics, structural and physiological. Is the X-chromosome typical in these relations? Shall we find similar effects in the other chromosomes? In addition to the X-chromosomes, there are the autosomes, which are present as a rule as several or many pairs. In man there are 23 pairs of autosomes: in Drosophila there are three pairs.

There is, further, in many species the single Y-chromosome. The autosomes and the Y-chromosome are distributed from parent to offspring in characteristic ways, differing from the distribution of X-chromosomes. If they affect characteristics, these differences should appear in a different system of inheritance. We shall examine first the autosomes.

The Autosomes: Typical Mendelian Inheritance

We have already seen that the autosomes play a role in development, in the fact that they help to determine the sex of the developing individual. In Drosophila, if the number of sets of autosomes, in relation to the number of X's, is changed, this changes the sex. Do autosomes also affect other characteristics? Do the different pairs of autosomes have different functions? Are there diverse types of autosomes, with different effects, in different individuals of a species, as is the case in relation to the X-chromosomes? Are there dominant and recessive effects of autosomes, as there are of X? If dominant and recessive characteristics depend on autosomes, then these characteristics must follow the distribution of the autosomes from parents to offspring, as the characteristics dependent on X follow its distribution. We must then examine the method of distribution of the autosomes whether there are characteristics that show this method.

Method of Distribution of the Autosomes

The distribution of the autosomes from parents to offspring is much simpler than that of the X-chromosomes.

The main facts are as follows:

- The autosomes are in pairs in all individuals. There are no individuals with unpaired autosomes, like the male of some organisms with relation to X.
- Thus the two sexes are alike as to their autosomes.
- In forming germ cells, one member of each pair of autosomes goes to each germ cell. The sperms and the ova are alike with respect to the autosomes.

  Thus the fundamental rule with relation to the autosomes is that each germ cell carries one member of each pair of autosomes.
- By union of two germ cells the pairs of autosomes are restored.
- Thus each one of the offspring carries in each pair of its autosomes one autosome from each parent.

Now consider a single pair of autosomes, such as the pair marked AA in. Suppose that in different individuals the members of this pair of autosomes produce different effects, as we have seen to be the case with the X-chromosomes of different individuals. And suppose that, as in the case of X, one of these effects is dominant, the other recessive. How would such characteristics be inherited? This question must be examined with care. For this purpose we may take as an example the different colours found in different varieties of peas, used in the classic experiments of Mendel. The yellow seed programmes of a certain variety is dominant, and results from the constitution of the autosomes of a certain pair. The green seed programmes of another variety is recessive, owing to a different constitution of the autosomes of that same pair.

Call the two autosomes of that pair in the dominant yellow race by the capital letters *AA*, while the two autosomes that give rise to the recessive green programmes in the other race may be called *aa*. The dominant parent may be designated *D*, the recessive parent *R*, When these two parents produce germ cells, each germ cell receives, of course, one chromosome for the pair. The germ cells from the dominant parent *D* have the autosome *A*, those from the recessive parent *R* have the autosome *a*. The germ cells *A*, from parent *D*, unite each with a germ cell a from parent *R* giving offspring (zygotes) containing both *A* and *a*. These first generation offspring are commonly designated *Fi* (first filial generation).

And since the characteristic carried by the autosome *A* is dominant, all these *Fi* offspring manifest the dominant characteristic *D* (yellow, in case yellow-seeded peas are crossed with green-seeded peas). Next observe the result of mating together two of these individuals *Aa*, of the *Fi* generation, giving the *Fa* (second filial generation).

Each parent *Aa* gives, according to the general rule, two kinds of germ cells, one kind carrying the dominant autosome *A*, the other the recessive autosome *a*. Each kind of germ cell from one parent unites with each kind from the other. That is, half the *A* germ cells from one parent unite with *A* germ cells from the other parent, half with a germ cells from the other parent; and the same is true for the a germ cells from the first parent.

This gives in equal proportions Fa offspring (zygotes) as follows:

$$AA + Aa + a\ A + aa$$

But as the two combinations *Aa* and *aA* contain the same chromosomes, the proportions may be written *AA* + 2*Aa-f-aa* The individuals of the constitution AA and Aa contain the dominant chromosome *A*; therefore they manifest the dominant characteristic, *D*. The individuals *aa*, having only the recessive chromosome *a*, manifest the recessive characteristic, *R*. Thus in this F^2 generation, so far as manifested characters are concerned, the proportion of offspring are 3 Dominant to *i* Recessive

Thus if dominant and recessive characteristics depend on the two members of a pair of autosomes, when a dominant individual *AA* is mated with a recessive individual *aa* the general result is bound to be that the immediate offspring (*Fi*) all manifest the dominant characteristic, while in the *Fa* generation (produced by mating two individuals from *Fi*) there will be three times as many dominant individuals as there are recessives. Now, this is the result that was discovered by Mendel in 1866; it formed the foundation for the working out of Mendelian heredity. The proportion 30 to *iR* in the *Fa* generation is commonly known as 'the Mendelian ratio'; it is one of the most characteristic proportions in this type of inheritance.

Mendel discovered that there are many different characteristics that yield this 3 to *i* ratio. If yellow and green peas are crossed, in the *Fa* generation there are 3 yellows to *i* green; if round peas and wrinkled peas are crossed, in F^2 there are 3 round to *i* wrinkled; and so for many other characteristics in the pea plants with which Mendel worked. And since the time of Mendel these same ratios have been found for thousands of different characteristics, in great numbers of different organisms, from the pea plant and *D* roso- phila to man. The fact that inheritance gives these ratios is commonly spoken of as Mendel's law, and the type of inheritance which it exemplifies is spoken of as Mendelian inheritance. The type that gives in the *Fa* generation 3*D* to *iR* is the commonest method of inheritance, much commoner than the sex-linked method, in which the characteristics depend on differences in the X-chromosomes. This is obviously because autosomes are more numerous than X-chromosomes.

We find then that inheritance according to Mendel's law takes place exactly as it would if the characteristics concerned depended on the two members of a pair of autosomes. As will be shown later, it can be proved experimentally that such characteristics do indeed depend upon autosomes. Thus the rules of Mendelism are the rules of distribution of the two members of a pair of autosomes, one having a recessive effect, the other a dominant effect. Mendelian heredity is heredity of characteristics that depend on differences in autosomes. Mendel did not know of the relation of the characteristics to the chromosomes. He discovered, however, certain other important facts about this type of heredity, facts which have become clearly explicable since it has been learned that this type depends on differences in autosomes. These facts must be examined. For this purpose it will be convenient first to define certain terms which are universally employed in dealing with inheritance.

By the union of two germ cells or gametes, as *A* and *a*, there is produced a zygote. If the two germ cells are alike with respect to the chromosomes or characteristics that they carry, the zygote so produced is spoken of as a homozygote; thus *AA* and *aa* are homozygotes. If the two germ cells that unite to form a zygote are unlike, as *A* and *a*, the zygote so produced is called a heterozygote, as *Aa*. The two corresponding characters, one dominant, the other recessive, that are produced by the two differing members of a pair of chromosomes are spoken of as alternative characters, or alleles, or allelomorphs. Any given individual manifests one or the other of the two alternatives, not both. For the alternative characters we shall employ henceforth the term alleles. The characteristic manifested by a zygote is its phaenotype while its actual constitution is its genotype. Thus, two zygotes *AA* and *Aa* both manifest the dominant character; the dominant character is the phaenotype, or the individuals are said to be phaenotypically dominant. But they are diverse in genotype; one is *AA*, the other *Aa*. In the case of a homo-zygote *AA* or *aa*, the phaenotype and genotype may be held to coincide. Mendel crossed yellow peas with green peas, and found, as before stated, that in the *Fi* generation all the peas were yellow, so that yellow is the dominant allelomorph. In the F^2 there were three times as many yellows as greens. The actual number of seeds in the F^2 generation was in his experiments 8023, of which 6022 were yellow while 2001 were green, so that the ratio of dominants to recessives was, accurately expressed, 3.01 to *i*.

Mendel found that if he bred further the individuals of the F^2 generation, some of these gave different results from others. In the pea plant it is possible to allow each plant to fertilize itself, so that both kinds of germ cells come from the same parent. When this was done with the individuals of the F^2 generation, the following results were observed: The green plants (recessive) gave only recessive green offspring (they 'bred true'). Of the yellow plants (dominant), one-third gave only yellow offspring (they 'bred true'), while two-thirds of them gave both yellow and green offspring, in the proportion of 3 yellows to *i* green. The reason for this is clear when we consider the chromosomes on which the characters depend. If *A* represents the chromosome (autosome) that gives the dominant yellow programmes, while a represents the chromosome that gives the recessive green, the original parents were *AA* and *aa*, the *Fi* generation were all *Aa*, while the *Fa* generation showed the proportions *AA*-*f* ² *Aa* + *aa*.

The recessive individuals *aa* are homozygotes; their two chromosomes are alike, being *aa*. When germ cells are produced by these homozygotic individuals, all these germ cells must contain the chromosome *a*, and when two of these unite in self-fertilization, of course they produce in all cases again individuals *aa*. These are, of course, recessives (green), like their parents; that is, the green parents aa 'breed true'. Similarly the dominant yellows, *AA* (constituting one-third of all the dominants), are homozygotes. Their germ cells all contain the chromosome *A*, and the offspring produced by union of such germ cells, in self-fertilization, are all *AA*, and so dominant yellow like the parents. One-third of all the dominants thus breed true. But the individuals *Aa* are heterozygotic; that is, they have two kinds of chromosomes, *A* and *a*, in the pair.

Thus they produce two kinds of germ cells, *A* and *a*, in equal numbers. And when these germ cells unite, they necessarily give, as we saw on an earlier page, offspring of three kinds, in the proportions *AA* + 2*Aa*4-*aa* Of these, *AA* and 2*Aa* will be dominants (yellow), while *aa* will be recessive (green). Thus the homozygotes *aa* and *AA* breed true, while the heterozygotes, *Aa*, do not. As Mendel discovered, if the heterozygotes *Aa* are mated with the recessive homozygotes *aa*, they give two kinds of offspring, *Aa* and *aa*; half (*Aa*) being dominant,the other half (*aa*) being recessive. If the heterozygotes *Aa* are mated with the dominant homozygotes *AA*, the offspring *AA* and *Aa* are all dominant, though half (*AA*) are homozygotes, the other half (*Aa*) heterozygotes.

All these results are typical for the descendants of two parents different in a single pair of chromosomes, one pair being dominant, the other recessive. For reference it will be well to make a table of these results, characteristic for Mendelian inheritance.

The Results of Mendelian Inheritance

Parents: *AA* and *aa*; both homozygotic. *Fi*: *Aa*; all alike, dominant heterozygotes. *Fs*: *AA* + 2*Aa* + *aa*. Three-fourths of the individuals dominant, onefourth recessive. Half the individuals homozygotes (*AA* and *aa*), the other half heterozygotes (*aAa*). Each type of homozygote bred by itself produces offspring like itself. The heterozygotes *Aa*, when bred by themselves, produce again offspring. *AA* + 2*Aa* + *aa Aa* mated with aa yields again *Aa* + *aa Aa* mated with *AA* yields again *Aa* + *AA*. All of these results are found to hold true for thousands of different characteristics.

We may therefore summarize the situation in the following general propositions:

- When two different individuals are crossed, in many cases the differences between them follow the same method of distribution as do the two members of a pair of autosomes ýÿone being dominant, the other recessive.
- This distribution is what is known as typical Mendelian heredity.
- The two sexes do not differ in these respects. Each sex has the same proportion of dominants and recessives.
- The rules of Mendelian inheritance are essentially the rules of distribution of the members of pairs of autosomes.

Characteristics that depend on two or more pairs of autosomes.We have thus far followed the distribution of characteristics that depend on differences in a single pair of autosomes. But most organisms have several or many pairs of autosomes: the fruit-fly has three pairs, man has twenty- three. Do some characteristics depend on differences in one pair of autosomes, others on differences in other pairs? Have the different pairs of autosomes thus different functions in development and inheritance? We find that the different pairs of autosomes have indeed different functions, and that some characteristics depend on one pair, some on another. This is shown in many cases in which two parents differ in two diverse characteristics. One characteristic may then act as if dependent on one pair of autosomes, the other on another pair. As an example of this, we may take a case studied by Mendel in peas.

One of the original pea plants had seeds that were (i) yellow, and (2) round, or smooth. These, as it turns out, are both dominant characters. The other plants, crossed with those just mentioned, had recessive characters; the seeds were (i) green, and (2) wrinkled. When these two kinds of pea plants were mated together, it turned out, as we shall see, that yellow and green behave in inheritances as if due to a difference in the two parents in the effects of a certain pair of autosomes; that is, they are alleles. ·

Round and wrinkled behave as if due to a difference between the two parents in the effects of another pair of autosomes. We may therefore represent the two pairs of autosomes. Representing the dominant characters by capitals, the recessives by lower-case letters, we may in the diagram represent the chromosomes giving origin to the four characters as follows:

A = yellow a = green

B = round b = wrinkled

The germ cells from these parents contain, of course, one member of each pair, as shown in Fig. One parent produces gametes all of the constitution *AB*, the other gametes all of the constitution *ab*. When a gamete *AB* from one parent unites with a gamete *ab* from the other, all the offspring (*Fi*) thus produced have the constitution *AaBa*, as shown at *Fi* in Fig.. Since the dominant autosomes *A* and *B* are both present, all these individuals are dominant for both pairs of characters; that is, they all have seeds that are yellow and round, as Mendel found. But all are heterozygotes-for both of these characters.

Next we mate together two of these *Fi* individuals to give the F^2 generation. In forming gametes the rule is, of course, that each gamete contains one member of each pair of chromosomes. But the two pairs, *Aa* and *Bb*, are distributed to the gametes quite independently. Thus, some get *A* and *B*, some *A* and *b*, some a and *B*, some *a* and *b*. There are thus from each parent four types of gametes, *AB, Ab, aB*, and *ab*, each type being represented by many gametes.

The gametes of each of the four types from one parent unite in approximately equal proportions with gametes of each of the four types from the other parent. That is, gametes of the type *AB* unite in different cases with *AB, Ab, aB* or *ab* from the other parent; and so of each of the other gametes of the first parent. The result is to give a set of 16 combinations of gametes (some of which are alike, however, as will be seen). It is not necessary to carry further in our diagram the outlines of the chromosomes.

Inheritance Depending on Two Pairs of Autosomes

AB AB AB AB

AB(i) *Ab*(6) *aB*(7) *ab*(s)

Ab Ab Ab Ab

AB(6) *Ab*(2) *aB*(5) *ab*(8)

aB aB aB aB

AB(7) *Ab*(5) *aB*(3) *ab*(9)

ab ab ab ab

AB(5) *Ab*(8) *aB*(9) *ab*(4)

But, as will be observed, not all these 16 are diverse; thus *ABAb* is the same in constitution or genotype as *AbAB*. If we number the different combinations, and give the same number to those that are alike, as is done (in parentheses) we find that there are nine different combinations or genotypes. That is, two parents that are both *ABab* produce offspring of nine different constitutions or genotypes.

When the parents differed in but one pair of characters there were but 3 different combinations. When there are two pairs the number is 3 > or 9⁻. These nine different combinations will manifest the dominant characters *A* or *B*, in all cases where an *A* or *B* is present whether present once, as *A*, or twice, as *AA*); they will manifest the recessive characters *a* or *b* in all cases in which *a* or *b* is present twice, no corresponding capital letter being present. Collecting together these different types, we find the different characteristics to be manifested in the following proportions of the zygotes:

$$9AB + 3Ab + saB + 1ab$$

That is, 9 out of 16 are dominant for both pairs of characters, 3 are dominant for *A*, recessive for *b*; 3 dominant for *B*, recessive for *a*, and *i* out of 16 is recessive for both. There are thus 4 different phaenotypes.

In Mendel's experiment with peas, in which the two pairs of characters were:

1. Yellow and green;
2. Round and wrinkled; in *Fi* the proportions were 9 that were yellow and round,
3. That were yellow and wrinkled,
4. That were green and round, and *i* that was green and wrinkled.

These proportions hold generally when there are two pairs of characters connected with different pairs of autosomes. It may be observed that the proportions of the individuals of different constitution or genotype in F^2, descended from a cross between parents that differ in respect to two pairs of autosomal characters (so that the parents are *AABB* and *aabb*), can be found directly from the proportions for a single pair of characters. In *Fa*, for the single pairs the proportions are *AA* + 2*Aa-haA* and *BB* + 2*Bb* + *bb* respectively. If now we multiply together algebraically these two polynomials, we obtain the relative proportions for the different combinations produced when the two pairs are together.

These proportions are the following:

AABB + 2*AaBB* + *aaBB* + 2*AABb* + 4*AaBb* +

(1) (7) (3) (6) (5)

2*aaBb* + *AAbb* + 2*Aabb* + *aabb*

(9) (2) (8) (4),

Certain other relations among the nine different types are important, since they hold for all cases of two pairs of characters that are dependent upon different pairs of autosomes.

These are the following:

- There are four combinations (Nos. 1, 2, 3, 4), constituting one-fourth of all the zygotes, that are homozygotes for both pairs of genes. The four homozygotes are diverse in their combinations and in the characters they manifest. Any one of these when self-fertilized or mated with another like itself breeds true: that is, it produces offspring that have the same constitution as the parents.
- Four out of the 16, or one-fourth of all the zygotes, are heterozygotes for both pairs of characters. They form the class numbered (5), and constitute the diagonal from upper right to lower left. Any of these when self-fertilized or mated with another like itself gives again the nine different types, in the proportions above set forth.
- The one-half of all the zygotes, are homozygotes for one of the two pairs, heterozygotes for the other. Four are homozygotic for *A* and heterozygotic for *B*; the other four are heterozygotic for *A* and homozygotic for *B*. In these 8 there are but 4 diverse combinations or genotypes.

If any one of these is self-fertilized, it yields 3 diverse types with respect to the pair for which it is heterozygotic. Thus, the class (6), having the constitution AABb, yields when self- fertilized three combinations in the following proportions:

AABB + 2*AABb* + *AAbb*

The above relations hold for all cases in which two pairs of characteristics are dependent on differences in two different pairs of autosomes, whatever the method of designating the characters or chromosomes. In a similar way, two parents may differ in three characters that are dependent on three different pairs of autosomes. Such parents could be represented as *AABBCC* and *aabbcc*.

Their progeny (*Fi* generation) would all have the constitution *AaBbCc*. Such *Fi* individuals of course produce a variety of germ cells, in accordance with the principles that each germ cell receives one member of each pair of autosomes, and that the different pairs of autosomes are distributed independently. Each parent *AaBbCc* thus yields the following 8 types of germ cells:

ABC, ABc, AbC, aBC, Abe, aBc, abC, abc.

When each of the 8 kinds from one parent unites with each of the 8 kinds from the other, there are produced, of course, 64 groups. If we classify the 64 individuals by the characters which they manifest, we find that there are 8 different phae- notypes manifesting respectively the following characteristics:

ABC, ABc, AbC, aBC, Abc, aBc, abC, and *abc*

That is, some are dominant for all three characters; some for a particular two and recessive for the other; some are dominant for a particular one and recessive for the other two; and some are recessive for all three. The different phaeno- types are present in different proportions; the proportions when a large number are examined are as follows:

2*ABC* + *gABc* + *gAbC* + *gaBC* + $*Abc* + *saBc* + 3*abC* + *iabc*

The total number of different combinations (genotypes) that occur in such a case is. These combinations can be obtained in their correct proportions by multiplying together algebraically the expressions: Similar but still more complex series are obtained if parents are interbred that differ in four or more characters depending on four or more different pairs of autosomes. If the number of pairs of autosomes is called ↔, the following relations hold:

If parents differ in n autosomes:

- The number of diverse types of gametes produced by the individuals of the *Fi* generation is 2^n.
- The number of different combinations (genotypes) in the zygotes of F^2, formed by mating the 2^n gametes from *Fi*, is s^w.
- The number of different phaenotypes produced in *Fa* is 2^n.

These relations yield very great numbers of diverse combinations, in cases where the number of pairs of autosomes is large. Thus in man, with 23 pairs of autosomes, if all the autosomes differed in the two parents, the number of types of germ cells produced by the *Fi* individuals would be 2, or many millions, the number of possible diverse genotypes in F^2 would be 3, or many billions, and the number of possible different phaenotypes would be 2. (As will appear later, the number of different possible combinations is greatly increased by the exchange of parts by the members of a pair of chromosomes.)

Thus typical Mendelian inheritance, dependent on differences between different pairs of autosomes, can be treated to a certain extent mathematically.

The individuals in F^2 and later generations occur in proportions represented by simple mathematical ratios.

It will be worth while to formulate the main facts that we have brought out as to autosomal inheritance:

- Some single pairs of characters, dominant and recessive, follow in inheritance the method of distribution of a single pair of autosomes.
- Some pairs of inherited characters follow the rules of distribution of two or more different pairs of autosomes, in the following particulars:
  - One member of each pair goes into each germ cell.
  - The members of the different pairs are distributed to the germ cells independently, so that all possible combinations containing one member of each pair are formed with equal frequency.
  - Each type of gamete from one parent unites with each type from the other parent, the different matings being equally frequent.
- These methods of distribution result, for the characters manifested in the Fs generation, in certain typical numerical proportions or ratios, known as Mendelian ratios.
- This method of inheritance is known as Mendelism, or Mendelian inheritance. The rules of Mendelism are the result of the rules of distribution of pairs of autosomes.
- Most characteristics of organisms are inherited in this way; many more than are inherited in the sex-linked manner. This is evidently because there is but a single pair of X-chromosomes, while there are several or many pairs of autosomes.
- Thus it is clear that, among the different individuals of a species, there are many different types of autosomes (as there are different types of X-chromosomes), producing many kinds of characteristics.

The T-Chromosome

There remains another chromosome, not included in the autosomes: namely, the Y-chromosome. Has this functions similar to those of X and of the autosomes? Does it too play a role in heredity? As we shall see, the Y-chromosome differs in many ways from the other chromosomes.

Let us first summarize the main facts as to the occurrence and distribution of Y-chromosomes.

- Some organisms have no Y-chromosomes. Among those are many insects the dog, the cat, the horse.
- In some organisms it is small and appears to be degenerate. This is the case in many insects; also in man.
- In some organisms the Y-chromosome is large, but of a different form from its mate X. Such is the case in Drosophila

- In some organisms Y is not distinguishable from X inform or size, but experiment shows that it differs from X in function.
- The Y-chromosome descends from father to son only, exclusively in the male line. Normally it never occurs in females.
- Hence any characters that depend on the presence of a particular kind of Y will *(a)* be found in males only, and will *(b)* never be inherited *through* females (since Y does not pass through females).

The T-Chromosome in Drosophila

The functions of the Y- chromosome have been more thoroughly studied in Drosophila than in any other organism. Most of the facts as to this have already been brought out, in connection with our account of the functions of the X-chromosome.

We may summarize as follows the known facts as to the functions of the Y-chromosome in Drosophila:

- The Y-chromosome, although it normally occurs in males, is not required for the production of male individuals. Through non-disjunction, male individuals are produced that contain no Y.
- But the presence of Y is necessary for the fertility of males.
- Y may exist in females, as a result of accidental nondisjunction; it does not affect their sex.
- During many years of intensive study no characteristics were found that follow the Y-chromosome.
- But recently it has been shown that the presence of Y prevents the appearance of a certain recessive character ('bobbed') that is due to a modified X-chromosome. This modified X results in the production of shortened bristles on the body of the fly. But this peculiarity does not occur if a Y is present, so that Y has the dominant normal allele for 'bobbed'. Indications of certain other ill-defined functions of the Y-chromosome have of late come to view.
- No diverse modifications or defects such as are known in great number for the X-chromosome are known in the Y-chromosome.
- Thus in Drosophila the Y-chromosome appears to have little function; it seems nearly but not quite inert.

Active T-chromosomes in certain Organisms; Relation to Male Secondary Sex Characters. In certain organisms the Y-chromosome is more active and varied in its effects. In such cases it shows in different individuals diverse modifications, comparable to the diverse types of the X-chromosome in Drosophila, and these produce diverse inherited characteristics that appear in males only. This matter is closely bound up with the question of what produces the diversity of secondary sex characters. In many animals there are special characteristics which, like the Y- chromosome, are found only in males.

These are the male secondary sex characters: horns, mane, beard, greater size, diversities in form, and the like. Since these appear only in the male, the question arises as to whether they may be due to the presence of the Y-chromosome.

With relation to this the following is the situation:

- In most cases, the presence of male secondary sex characters does not depend on the presence of the Y-chromosomes.

  This is shown by the following:

  — There are animals which have no Y-chromosome, yet have male secondary sex characters: for example, the dog, the cat, the horse; also various insects.

  — In Drosophila, males that contain no Y (as a result of non-disjunction) have the typical male secondary sex characters.

  — In some breeds of sheep the male has horns, while the female has none. It could be suggested that this is due to the fact that the male carries the Y-chromosome, while the female does not. But in other breeds, as Dorsets, both sexes have horns, although only the male has the T-chromosome. And in still other breeds, as Suffolks, neither sex has horns, although here as elsewhere the males have the Y-chromosome.

  If a female of the horned breed (Dorset) is crossed with a male of the hornless breed (Suffolk), in the Fi descendants the males have horns, although they receive their Y-chromosome from the hornless breed. They have inherited the horns through the mother. Thus it is clear that in these cases the presence of horns is not due to the presence of the Y-chromosome. Extensive work on the inheritance of horns and horn- lessness in sheep, goats and cattle consistently gives results which show that the presence of horns is not due to the presence of the Y-chromosome, or of any special type of Y-chromosome. In all these cases (a), (b) and (c) the presence of distinctive male secondary sex characters depends in some way on the fact that the male carries but one X-chromosome, and consequently produces the male hormone. Their absence from the female results from the fact that she carries two X-chromosomes, and consequently produces the female hormone.

- But in certain fish, some of the male secondary sex characters apparently do depend on the presence of a Y of a particular type, for they occur only where that type of Y is present, and they can therefore not be inherited through the mother. That is, the Y-chromosomes of certain individuals or breeds are so modified as to produce certain definite characteristics in the individuals that contain them, while individuals that contain another type of Y do not show these characteristics.

An example of this is found in the small fish Lebistes. In a certain race, the males have a black spot at the base of the dorsal fin, while the females are without the spot. In another race of the same species neither the males nor the females have the black spot. Thus the females of the two races are alike in lacking the black spot, while the males differ. Is this difference the result of differences in the Y-chromosomes of the two kinds of males? This may be tested by crossing members of the two races. In the spotted race, represent the X and Y-chromosome by capital letters; while in the non-spotted race, represent them by the small letters *x* and *y*.

A female of the spotted race, XX, is mated with a male of the non-spotted race, xy, thus:

- Females Males
- Parents XX *xy*
- Non-spotted
- *Fi Xx Xy*
- Non-spotted

The Y-Chromosomes

This produces in Fi females Xx, and males Xy. Neither contains the Y from the spotted race, and neither is spotted. The spot is not inherited through the females of the spotted race.

Next mate the female of the non-spotted race, xx, with a male from the spotted race, XY, as follows:

- Females Males
- Parents xx XY
- Spotted
- Fi xX xY
- Spotted

This yields in Fi females Xx, and males xY. These males carry the Y from the spotted race, and they are spotted. Thus far the spot goes with the Y from the spotted race. This Y may be traced further, for example, by breeding together the non-spotted females xx with the Fi males xY, from the preceding cross, thus:

- Females Males
- Parents xx xY
- Spotted
- F2 xx xY
- Spotted

This gives in the Fs femalcs xx, and males xY. These males are again spotted; they carry the Y from the spotted race. Wherever this Y from the spotted race is found, the black spot appears; wherever the Y is from the non-spotted race it does not appear. It appears clear therefore that whether the black spot appears in the male depends upon the type of Y- chromosome that it carries. There are two types of Y-chromo- some in the species, one producing the black spot, the other not producing it. It has been shown by Winge that there are many other programmes markings in this fish that depend for their occurrence on the presence of a particular type of Y-chromosome. They occur only in the males, and only in males that have a certain type of Y. In certain other fish also the Y-chromosomes have been shown to be diverse in different individuals, giving rise to distinctive characteristics in the male. In a certain plant, Melandrium, also, there are inherited characteristics that depend upon the presence of a particular type of Y-chromosome.

There are some indications also that in man the nature of the Y-chromosome affects certain characteristics. In a family described by Schofield, the male parent in the first observed generation had webbed toes. This was inherited in the male lines only by all the males for four successive generations. It thus followed exactly the distribution of the Y-chromosome from the original male parent. It seems probable therefore that it is due to a defect or modification in the Y-chromosome of that individual. Many other cases of webbed toes or fingers are known that occur in both sexes, so that they are clearly not due to the Y-chromosome.

On the whole, the situation as to the Y-chromosome and its relation to inheritance may be summarized as follows:

- In most organisms Y is a degenerate chromosome, lacking most of the functions manifested by the other chromosomes.
- In some organisms degeneration has gone so far that the Y-chromosome has completely disappeared.
- In other cases it exists, but is very small.
- But in some species the Y-chromosome has effects on hereditary characters that are similar to those produced by other chromosomes. Different types of inherited characters result in such cases from diversely modified Y-chromosomes.
- Characteristics dependent on the presence of a particular type of Y-chromosome are inherited in the male line exclusively.

Three Main Methods of Distribution of Inherited Characteristics

There exist therefore three diverse systems of distributing hereditary characteristics, depending respectively on the X- chromosomes, the autosomes, and the Y-chromosomes. In each of these the rules of inheritance are given by the special method of distribution of the three types of chromosomes, taken in connection with the facts of dominance and recessiveness. Inheritance which follows the distribution of the X-chromosomes is commonly called sex-linked inheritance. That following the distribution of autosomes is known as autosomal or typical Mendelian inheritance. Less important than those just named is Y-chromosome inheritance. All these methods are included under the term Mendelian inheritance, employed in a broad sense. In some organisms still other methods of inheritance exist. In higher animals and plants most inheritance occurs in one of the three methods named in the preceding paragraph.

Characteristics Affected both by Genes and by Environment

Many of the same features that can be altered by changing the genes can likewise be altered by appropriate changes in the environment. Characteristics do not fall into two classes, one exclusively hereditary (or dependent on genes), the other exclusively environmental; but any characteristic is affected both by the materials of which the organism is composed, and by the action of the conditions on these materials. Yet, as we shall see, in practice some characteristics are more readily altered by environmental conditions than are others. Certain characteristics owe most of their peculiarities to diversities among genes. Others are readily affected both by genes and by environment. Still others depend mainly on environmental conditions. Particularly illuminating are the cases in which characteristics depend in marked degree both on genes and on environmental conditions, so that they are altered by changes in either. A number of well-known cases of this kind will be presented, selecting those that are from a technical point of view fully known. In that animal whose genetics is best known, the fruit-fly Drosophila, many of the characteristics that have been studied are defects or abnormalities which are sharply dependent on the presence of particular genes. Such a one is abnormal abdomen. This appears in the fact that the abdomen is ill formed, the segments not being regular nor sharply marked off. This abnormality is found to be due to a defect in a certain gene of the X-chromosome, so that it shows sex-linked inheritance, the abnormality being dominant. Unless that particular defective gene is present the abdomen is normal. But it is dependent too on the environment.

Individuals with the abnormal gene, if developed in a moist atmosphere, have the abnormality. But if those same individuals are developed in a dry atmosphere, they have normal abdomens. To produce the abnormal abdomen, we must have first the particular kind of defective gene that yields it. But second, even if that is present, we must have, too, the moist atmosphere for development; without it the abnormality is not produced, whatever the gene. Another character in the fruit-fly that acts in a similar way is what is called reduplicated legs. If a certain defective gene is present in the X-chromosome, the animals show a tendency to produce double legs, or even triple or multiple legs, in place of single ones. But this does not happen unless the young develop at a low temperature. To produce reduplicated legs, then, there must be present a certain type of gene, and also a certain external condition. Unless both are present, the peculiarity is not produced. It is clear from these cases that even though an individual inherits the type of gene necessary to produce a certain defect, it may not be inevitable that he should have that defect. A special environment may correct what inheritance leaves imperfect. These considerations apply to many things in man as well as in other organisms.

Another example of the fact that what is produced in development depends both on the genes and on the environment is seen in the production of giants in the fruit-fly. A modified gene, located in the X-chromosome, near its left end, causes the animals (if they have no other type of X-chromo- somes than this) to become giants; they grow to nearly twice the size of the ordinary fly. But this increase in size takes place only if the animals having the modified gene are well fed during a certain period of their early lives. If not well fed at this particular time in life, they grow no larger than the usual flies. The giant size requires to produce it a particular type of environment acting on a particular type of gene. If either condition is not fulfilled, giants are not produced. Again, in plants, the difference between green plants, containing chlorophyll, and white ones, is in some cases due to a difference in genes. Plants with a certain kind of genes will remain white even though grown in the light. In other cases, the difference between white and green plants is not the result of a difference in genes, but of a difference in environment. Plants that live in the dark remain white, even though they contain genes that can produce chlorophyll, while similar plants grown in the light are green. That is, to produce chlorophyll, a certain type of genes is necessary, and also a certain type of environment. Again, there are red and white varieties of primroses; the difference in programmes is inherited; it is due to gene diversities. But a certain variety produces red flowers when grown in a cool place, white flowers when grown in a warm, moist region, as in a greenhouse. The same diversity in programmes is now due to an environmental difference.

Complex interrelations of genetic and environmental action have been described by Emerson with relation to diverse colours in maize plants. In different types, the leaves, the husks, the flowers, and other parts, differ in programmes. Emerson describes and Figs six main types, under each of which are subtypes.

These are the following:

- Purple, with its subtype, weak purple.
- Sun-red, with a subtype, weak sun-red.
- Dilute purple.
- Dilute sun-red.
- Brown.
- Green, with a number of subtypes.

The interrelations of genetic and environmental factors in these colours may be summarized as follows:

- Each programmes and each subtype is constant and hereditary, when the different types are bred under like conditions.
- Each of the colours depends on many different genes, like the eye programmes in Drosophila; they may be altered by changes in any of these genes. The different colours result from different combinations of the genes.
- When the different types are crossed they give typical Mendelian inheritance: in some cases with simple ratios, in others with complex ratios. Thus far the conditions are those for typical hereditary characters; the differences are the result of gene differences, and they follow the distribution of the different types of genes in inheritance. But, also, the colours depend on the conditions to which the plants are subjected while developing. Change of certain conditions changes the colours, as does alteration of the genes. Some of the relations are the following:
- The 'sun-red' type requires exposure to the sun for production of the red programmes. Parts protected from the sun are green. In the 'green types, on the other hand, subjection to the sun does not cause the production of red programmes. Thus the difference between red and green plants is due in some cases to diversity of genes, as when 'sun-red and 'green types are both grown in the sun. In other cases the difference between red and green plants is due to environmental difference, as when 'sun-red plants grown in the sun are compared with those of the same type grown in the shade. The plants of the 'purple' type do not depend on exposure to the sun for their programmes; they are purplish in programmes whether grown in sun or shade. The degree of programmes produced by exposure to the sun depends on what genes are present. In the typical 'sun-red' variety, exposure to the sun causes a red programmes widely distributed on the plant. In the 'dilute sun-red', exposure to the sun under similar conditions causes only a little red, at the bottom and tips of the leaves, while in the 'green type no red appears. The nature of the soil affects the programmes. 'Dilute sun-red' grown in poor soil and subjected to the sun has almost all parts red; in good soil, with the same relation to the sun, it has very little red. The 'green and 'brown' types are not influenced in their programmes by the nature of the soil.
- Certain particular chemicals in the soil affect the programmes. Presence of much nitrogenous matter tends to suppress the red; if little is present the red programmes is marked.
- Storage of carbohydrates in the leaves of the programmes-producing varieties causes increase of the red programmes. The amount of

carbohydrate in the leaves may be increased by experimental procedures, as by removing the growing ears, or by bending a leaf so as to break the conducting vessels; the leaves thereupon become red in the parts with increased carbohydrate. But this effect is not produced in the 'green' types; they do not produce the red programmes even in presence of stored carbohydrates.

It appears clear, therefore, that in maize different genes cause the different varieties to have different metabolic processes, different chemical reactions, some of which cause the production of colours, while others do not. Also, diverse environmental conditions likewise cause changes in these metabolic processes, inducing programmes production in some cases.Whether programmes occurs, and its type and amount, depend on interaction of genetic and environmental conditions. Another instructive case of the interaction of genes with environmental conditions is furnished by the conditions determining the number of facets in the compound eye of Drosophila. In the normal eye there are present about 800 facets. In certain of the flies a gene at locus 57 in the X-chromosome is so modified as to cause a great reduction in the number of fully developed facets. In this condition, known as Bar-eye, there are only about 80 fully developed facets: these form a 'bar' across the eye. This Bar-eye condition is dominant; the number of facets is reduced even though one of the two genes of the pair is normal. We may call the modified gene that causes the reduction in the number of facets the Bar gene.

The number of fully formed facets, it is found, depends on whether there is present but one or two of the Bar genes, and also on whether there is present in addition a normal gene. In the males (having of course but one X), the presence of the Bar gene reduces the number of normal facets to about 90. In the female, if both the X's contain a Bar gene, the number of facets is about 65. If the female is heterozygotic, so that but one of her X's contains the Bar gene, while the other is normal, the mean number of facets is about 358. Finally, in either male or female, if no Bar gene is present, the number of facets is about 800. Later it was discovered that this Bar gene is sometimes modified in a different way, so as to cause a less marked reduction in the number of facets. This modification is known as infra-Bar. A female containing but one of these infra-Bar genes has about 716 facets; if she contains two she has about 348. Again, by an extraordinary process, almost or quite unique in known genetics, in some cases two of the Bar genes get united by crossing-over into one X-chromosome. This condition is known as double Bar. When both of the chromosomes carry this double Bar, the number of facets is but about 25. If one carries double Bar, the other single Bar, the number of facets is about 36. Thus the number of facets in the eye depends in many ways on the number and conditions of the genes present. This may be shown in the following diagram, representing the conditions in the two X (or Y) chromosomes of any individual, with the corresponding numbers of eye facets present. In this diagram + = the normal, unmodified X-chromosome Y = the Y-chromosome of the male *B* = the Bar gene *iB* = the infra-Bar gene *BB* = the double Bar condition (two Bar genes in one chromosome) Ghr. *i*: + *iB B iB B iB B BB BB BB BB* Chr.s: + + + *iB* Y *BB* + *iB B BB* Facets 800 716 358 348 98 73 65 45 42 36 25. All these various conditions are inherited as unit-difference, sex-linked characters; they are typical hereditary conditions.

But it is found further that the number of facets in most of these conditions depends also on the temperature to which the developing animals are subjected. The numbers given above are those observed by investigators working with fruit- flies cultivated at ordinary, not precisely controlled, temperatures. If different sets of flies are kept at different temperatures, it is found that high temperatures reduce the number of facets in the individuals that carry the Bar genes, though not in the normal individuals. Krafka found the following mean numbers of facets in the various gene conditions, at different temperatures:

Females Degrees	Bar Males B-B	Bar Females B-Y	Double Bar Males BB-BB	Double Bar BB-Y	Temperature,
J5	214	270	5i	61	
20	122	161	33	37	
25	81	121	25′	28	
300131	40	74	‘5	‘4	

Thus the number of eye facets depends both on the genes and on the temperature. The number 61 may be produced either by a double Bar Male (*BB*–Y) at 15 degrees or by a Bar Female (*B*–*B*) at between 25 and 30 degrees. Changing either the genes or the temperature will change the number of facets. Similar situations occur in vertebrates; one of these may be described briefly. The spotted salamander, Salamandra maculosa, is largely black in programmes, with spots or stripes of yellow.

The amount of yellow differs in different cases, so that on the whole some specimens are lighter, some darker. In the lighter individuals the yellow spots tend to run together into rows. The different grades are heritable; certain stocks are regularly darker, others lighter. If darker and lighter individuals are mated, the two conditions show Mendelian inheritance, the darker condition being dominant all the Fi generation are darker, while in Fa there are three dark to one light.

The difference between the two thus depends on a difference in one gene. But it can be shown also that the extent of programmes depends on the environment. If the salamander is kept continuously for a long time (a year or two) on a light background, the yellow spots increase in size: they may run together into stripes or into large yellow areas *B*, so that the animals become much lighter. If, on the other hand, they are kept for long periods on a dark background the yellow areas decrease in size; the animals become on the whole darker. The extent of the yellow programmes thus depends both on the genes and on the environment; it can be altered by appropriately changing either.

In all these cases the following general principles apply:

- What are really inherited (passed on from parent to offspring) are the genes; that is, certain materials that in certain combinations and under certain conditions give rise to certain definite characteristics.
- With the same original genes, different environmental conditions may induce the production of diverse characteristics.
- Also, with the same environmental conditions, different genes may induce the production of different characteristics.
- The same difference in characteristics that is in one case produced by diversity of genes is in other cases produced by diversity of environment.

There can be little doubt that these general statements apply to many characteristics in man. Some combinations of human genes form an individual that is a much better culture medium for the bacteria of tuberculosis than are others. A person who gets such a combination of genes will develop tuberculosis, if he lives in a region in which the germs of that disease are abundant; while another person, with a different set of genes, will not be subject to tuberculosis, even though living under the same conditions. And the person with the genes that make him susceptible to tuberculosis will not have the disease if he prevents infection by the tubercle bacillus.

In relation to such susceptibility there are doubtless a great number of grades, dependent on just what combination of genes is present. Some are extremely susceptible, others less so but still prone to the disease, and so on, up to individuals whose gene combinations are such as to give them natural immunity to tuberculosis. What then shall be answered when it is asked whether tuberculosis is hereditary or not? One can only say that some gene combinations predispose to it more than others, so that a hereditary factor is involved. But also there is a necessary environmental factor; it is the interrelation of the two sets of factors that determines whether the individual shall be afflicted with the disease.

Similar considerations apply to any other disease, acute or chronic. Certain combinations of genes are doubtless more readily attacked by plague, by smallpox, by typhoid, by pneumonia, than others, just as certain genetic constitutions yield more readily to extremes of temperature, to exposure to the elements, to unfit food. There is no affair of life in which the gene combination borne by the individual does not play a part. But the environment too plays a part, often an overwhelmingly important part. The relative role played by genes on the one hand, and by environment on the other, is not the same for all characteristics. Differences in certain features of the organism are mainly due to differences in the genes of which they are composed. In other characteristics, differences are due mainly to environmental diversities. There are all sorts of intermediate conditions between these extremes. We may pass in review a number of different types of characters from this point of view.

Most of the diversities in obvious physical features in most animals are the result of differences in the genes, so that they are mainly matters of inheritance rather than of environment. In Drosophila the forms of the parts of the body, of the legs and wings, the distribution of bristles, the venation of the wings, the colours of body and eyes, are all changeable through the action of diverse genes, but diversities in these features due to environmental action are relatively rare. This is the case also for such characters in the vertebrates, including man. Most of the physical characters of human beings form of their features, complexion, programmes of hair, form of limbs, and the like, as well as sex are settled mainly while the child is carried in the mother's body, when the differences in environment are very slight.

The diversities between individuals in these respects are almost entirely though not quite entirely the result of differences in the genes; they are matters mainly of inheritance. This is demonstrated by what we find in twins. The one-egg twins, having the same genes, are usually closely alike in their physical peculiarities, while the twins derived from different eggs, and so having different genes, are much more diverse with respect to such physical characteristics. At times, of course, one-egg twins become diverse through accidents or diseases which affect one and not the other. But physical diversities in twins due to genetic differences are much more frequent than those due to environmental differences.

In some organisms the gross physical features are readily altered by environmental diversities, so that differences between individuals resulting from the different conditions under which they have lived are common. Such is the case with the habit of growth in higher plants; trees and other plants grown under different conditions are very different in form. In some of the lower animals, such as the hydroids, a similar situation is found. It is obvious that the details of behaviour are in a high degree under the influence of environmental conditions. The particular acts performed at given moments in any given organism are determined almost entirely by the conditions under which it finds itself. Yet organisms of different genetic constitution behave very differently under effectively the same conditions, and the general pattern of the behaviour of organisms may be largely determined by the genetic constitution.

Bibliography

Annie Tan: *Blueprints Notes and Cases - Biochemistry, Genetics and Embryology*, Blackwell Science, Delhi, 2004.

Balkwill, Fran: *DNA Is Here to Stay*, Carolrhoda Books, Inc., Minneapolis, 1993.

Banerjee, S : *Cell Biochemistry*, Dominant Pub, Delhi, 2009.

Bhatia, A L : *Biochemistry and Endocrinology*, Indus Valley, Delhi, 2002.

Brookes, Martin: *Get a Grip on Genetics, Time Life Books*, East Sussex, England, 1998.

Bruce, Donald : *Engineering Genesis: The Ethics of Genetic Engineering in Non-Human Animals*, London: Earthscan, 1998.

Cahill, Lisa: *Genetics, Theology, and Ethics: An Interdisciplinary Conversation*, New York: Crossroad, 2005.

Curtis, M.: *The Geometry of DNA: A Structural Revision*. London: Blue Gallery, 1996.

Diaz, E.: *Microbial Biodegradation: Genomics and Molecular Biology,* Caister Academic Press, UK, 2008.

Drlica, Karl : *Understanding DNA and Gene Cloning: A Guide for the Curious*, New York: Wiley, 1997.

Fukuyama F.: *Our Posthuman Future: Consequences of the Biotechnology Revolution,* Farrar, Straus and Giroux, NewYork, 2002.

Garza-Valdes, L.A.: *The DNA of God?* Hodder and Stoughton, London, 1998.

Hames, B. D.: *Genes and Embryos*, Oxford University Press, New York, 1989.

Jonoska, N.: *Self-Assembling DNA Graphs, DNA-Based Computers VIII*, Springer-Verlag, Berlin, 2003.

Kallenbach, N.R.: *Molecular Structure: Chemical Reactivity and Biological Activity,* Oxford University Press, Oxford, 1988.

Lewis, Ricki: *Human Genetics: Concepts and Applications,* Boston, Mass: McGraw Hill. 1994.

Mayr, E.: *Change of Genetic Environment and Evolution*, Allen and Unwin, London, 1954.

Nisha Khalsa: *Essentials of Biochemistry*, Aavishkar Pub, Delhi, 2008.

Old, R.W.: *Principles of Gene Manipulation*, London, Blackwell Scientific Publications, 1989.

Rabus, R: *Microbial Biodegradation: Genomics and Molecular Biology*, Caister Academic Press, UK, 2008.

Sturtevant, Alfred: *History of Genetics*, Harper and Row, New York, 1965.

Stwertka, Eve : *Genetic Engineering*, Franklin Watts, New York, 1982.

Stwertka, Eve and Albert: *Genetic Engineering*, Franklin Watts, New York, 1982.

Vasudev Bhatnagar: *Cell Science and Technology*, Campus Books, Delhi, 2009.

William, Kovacs D.: *An Introduction to Geotechnical Engineering*, Prentice Hall, UK, 1981.

Williams, J.G. and R.K. Patient : *Genetic Engineering*. Oxford, England: IRL Press, 1988.

Wilson, I.: *The Blood and The Shroud*. London: Weidenfeld & Nicolson, 1998.

Winston, R.: *The Future of Genetic Manipulation*, Phoenix, London, 1997.

Index

❑❑❑